The Politics and Possibilities of Self-Tracking Technology

The Politics and Possibilities of Self-Tracking Technology: Data, Bodies and Design

BY

SUNEEL JETHANI

University of Technology Sydney, Australia

emerald PUBLISHING

United Kingdom – North America – Japan – India – Malaysia – China

Emerald Publishing Limited
Howard House, Wagon Lane, Bingley BD16 1WA, UK

First edition 2021

British Library Cataloguing in Publication Data
A catalogue record for this book is available from the British Library

ISBN: 978-1-80043-339-7 (Print)
ISBN: 978-1-80043-338-0 (Online)
ISBN: 978-1-80043-340-3 (Epub)

Printed and bound by CPI Group (UK) Ltd, Croydon, CR0 4YY

ISOQAR REGISTERED
Certificate Number 1985
ISO 14001

ISOQAR certified Management System, awarded to Emerald for adherence to Environmental standard ISO 14001:2004.

INVESTOR IN PEOPLE

Table of Contents

Dedicated to the memory of Keshav Kumar Jethani

About the Author

Suneel Jethani is a Lecturer in Digital and Social Media at the University of Technology, Sydney. He has published work in *International Communication Gazette*, *Communication, Politics & Culture*, *Continuum* and *M/C Journal*.

Acknowledgements

First and foremost, I want to acknowledge my mother Pushpa and sister Nalani Jethani and thank them for their unwavering support and care. I would like to acknowledge and thank so many of my friends, colleagues, collaborators and mentors that it's almost impossible to do here. To Scott McQuire and Robert Hassan thank you for your mentorship, guidance and constructive feedback over the many forms that this project has undertaken. To Mathan Ratinam, thank you for saying something that started me along this path many years ago. To Thibault Behan, Jo Eaton Warren Davies, Adam Lodders and Helen Grogan thanks for your friendship, advice and for all those times you randomly checked in on me while I was finishing the manuscript. To my friends, colleagues and collaborators: Dale Leorke, Luke van Ryn, Belinda Barnet, Robbie Fordyce, Luke Heemsbergen, Rowan Wilken, Anthony McCosker, Bjørn Nansen, Tom Apperley, Chris O'Neill, Kate Mannell, Jasmin Pfefferkorn, Caitlin McGrane, Fan Yang, Wendi Li, César Albarrán-Torres, Kath Albury, Larissa Hjorth, Darshana Jayemanne, Andrew Iliadis, Isabel Pedersen, Sophie Freeman, Lyell Durkin, Emily van der Nagel, Dan Golding, Trent Griffiths, Samuel Kininmonth, Danny Butt, Ruth DeSouza, Debris Facility, Chris Marmo, James Meese and Sharee Gregory. All of you in some way have helped me shape the ideas in this book and helped me grow as a scholar. This book was written in the middle of a global pandemic and during a move from Melbourne to Sydney. To my new colleagues at University of Technology Sydney: Heather Ford, Amelia Johns, Paul Byron, Ben Abraham, Francesco Bailo, Tisha Dejmanee, Natalie Krikowa, Bhuva Narayan and Alan McKee thank you for providing a supportive and afflictive work environment, especially while this book was in its final stages. To Paul Stevens, and, Jen McCall, Kimberly Chadwick, Ramya Murali and the team at Emerald, thank you for your careful and attentive stewardship of this project.

Abstract

Collecting data about our lives, our bodies and our behaviours has become a part of everyday practice that promises greater self-awareness, healthier living and increased productivity. This book focuses on the dialectical relationship between those that design and use self-tracking technology in order to examine how logics of datafication redefine the body. It explores what these emerging relations mean for imagining, designing and analysing the sociotechnical systems that bring about self-tracking. The book provides a genealogy of self-tracking to situate the notion of quantified and quantifiable selves as problematic data regimes within contemporary digital culture. It charts the origins of self-tracking from within the blueprint of the Californian Ideology to a global social movement which now reaches beyond self-experimentation to encompass the wider trajectories of using wearable sensor technology in the neoliberal management of health, wellbeing and productivity. The book reframes and theorises the quantified self by re-examining and developing arguments of bodies which 'disappear' (Jewson) into, are made 'docile' (Foucault) by and get caught up in the 'rhythms' (Lefebvre) of datafication. The concept of a 'quantised' self is introduced as a means of reading into and exposing the inherent political interests being served when self-tracking technology is introduced into clinical, home and workplace settings. Drawing on the case studies of self-tracking in practice that precede, it the final chapter sketches the outlines of a mutual praxis of critique and design that facilitates the (re)imagination of the politics that are embedded into sociotechnical systems of self-tracking and considers possibilities of intervention.

Introduction

> In order to analyse a little more precisely this new technology of government, I think it is best to catch it the three major forms that any technology is able to take in its development and history: as a dream or, better, as a utopia; then as a practice as rules for real institutions; and then as an academic discipline. (Foucault, 1982, p. 154)

In *Do Artifacts Have Politics?* (1980), Langdon Winner argued that a critical understanding of technology is established not only through considering how useful an artifact is in solving a problem, but by thinking about the ways that technologies bring about the embodiment of power relations embedded into them as they are designed, implemented and studied. Technologies that are used to capture data on our health, behaviours, routines and habits are increasingly becoming a feature of everyday life. The past decade has the 'quantified self' emerge as an idea that represents self-mastery, better health, resilience, self-acceptance and greater personal agency. Underlying the notion of a quantified self is an evolving set of technologies and cultural practices called 'self-tracking'. In its most common form, self-tracking can be understood as the use of wearable or mobile technologies embedded with digital sensors that measure and track locations, heart rates, activity levels, sleep quality, speech patterns, concentration, mood, posture and a number of other variables. According to BCC Research, the self-tracking technology market which also includes mobile digital technologies for personal consumption such as wearable devices and the platforms developed by Google, Xiaomi, Garmin, FitBit Samsung, Apple, Microsoft and others is set to reach 71.9b USD by 2022 (BCC Research, 2017). The enthusiasm for self-tracking is well captured in a report on Smart Cities authored by the multinational consulting firm, Deloitte where the quantified self is described as a trend that:

> …uses sensor technology innovation embedded in wearable devices to […] allow people to monitor their physical conditions, from blood pressure to heart rate to blood glucose levels and adjust their behaviour in real-time to make better health decisions. Self-tracking data may make individuals more receptive to behavioural nudges and can be used by policy makers to reward healthy [and active] lifestyles. (Deloitte, 2019, p. 39)

The Politics and Possibilities of Self-Tracking Technology, 1–10

Published under exclusive license by Emerald Publishing Limited
doi:10.1108/978-1-80043-338-020211009

bring self-tracking technologies into the spaces and rhythms of everyday life. In taking an emphasis on the relations between use and design, in this book, I pick up on the call for greater conceptual work in critical data studies (boyd & Crawford, 2012; Dalton, Taylor, & Thatcher, 2016; Iliadis & Russo, 2016; Michael & Lupton, 2016) where self-tracking offers a thick and nuanced account of how the ongoing collection of personal data contributes to the production of subjectivity. There is little research addressing the materiality of self-tracking systems that engages specifically with design implications except Dolejsova et al. (2017), McKillop, Mamykina, and Elhadad (2018), Ayobi, Sonne, Marshall, and Cox (2018), and Schüll (2016) and in this work arguments tend to focus on self-experimentation as a dominant pattern of use. Within this frame of reference, design challenges are associated with issues such as data validity and reliability which are framed in terms of the collection standards required to meet the requirements of rigorous scientific research. This way of thinking about the design of self-tracking systems downplays concerns about privacy and reuse and runs the risk of inadequately acknowledging the tendency for obfuscation, falsification and avoidance which are not only an inherent part of the patterns of use we see in the use of many technologies that are designed to capture data, but also a form of 'strategic agency'(Bossewitch & Sinnreich, 2013) and 'informational self-defence' (Nissenbaum & Brunton, 2013, pp. 171–172).

If the question, posed by Marketa Dolejsova et al. (2017) and echoed recently by Wolf and De Groot (2020) of whether self-tracking systems should be designed with the aim of professionalising amateur self-experimentation, or to promote an agnostic approach to design that does not assume the superiority of one form of knowledge over another is to be addressed then several things need to be considered. This is a question that trades off adherence to narrowly defined technical logics against more open and genuinely human centric approaches to design. An extension of this is the question is: how are self-tracking systems are being designed in ways that enable access and reuse of data by others and to what end? Should the idea that sharing self-tracking data is altruistic and virtuous be taken at face value (by both users and designers)? How can critical frameworks developed with design in mind that help balance altruistic intentions with market led goals and objectives that support neoliberal agendas? Beyond this: how can thinking about self-tracking system design support outcomes that are socially responsible and how can they be brought closer to principles of inclusive design? Can those who are encircled by the logics of self-tracking systems be thought of as quantified, quantifiable and *quantised* selves? How can negative outcomes and other obtuse, idiosyncratic and iatrogenic aspects of self-tracking be genuinely accounted for within existing production methodologies such as Agile Methodology – which instructs designers to iterate rapidly, fail fast and build minimally viable products that can be refined once out to market. Similarly, Human Centred Design – which, in principle, tells us to be deeply user centric and inclusive of a broad range of archetypal user personas?

These are complex questions that I am attempting to address in this book. These questions require an approach that engages with a number of the inherent tensions that flow through this particular class of technology and sociotechnical system. First, is the tension between efficient data capture and the way that the logic of data capture mechanisms align to individual experiences that shape what it exactly is that makes self-tracking meaningful to users. Second, because sustained user engagement is critical to the success or failure of self-tracking technologies, their design might not include 'dimensions that are not a priori important to the ultimate goal of the tool' (McKillop et al., 2018, p. 10) and this has a limiting effect on how these tools might be granting personal agency or be socially progressive. Third, negative experiences of self-tracking might only be considered as part of the problem that a technology is trying to solve and not something that the technology itself brings about in the problem-space that it is being designed for. Fourth, data capture might not always be conceived as something that has varying resolution relative to the diverse range of times, spaces and bodies implicated in its use. Fifth, is the question of how contemporary examples of self-tracking technology draw from, or a distanced from, both materially and semantically their functional, morphological and ideological antecedents? Sixth, and finally, there is the relation between the device and the data it produces – if artifacts are political then which one should we be more concerned about devices or datasets? And by extension, to what extent is self-tracking data *personally* meaningful (Ayobi, 2018) and can design accommodate combining, switching or differentiating between devices and modes of data capture such as private, pushed, communal, imposed, and exploited (Lupton, 2016b) when there is a 'gap between the user and their data' (Schüll, 2016, p. 10)? Is this space where a politics of self-tracking emerges?

The approach I have taken in addressing these issues and questions is one that looks towards the goals, intentions and consequences of self-tracking. I do this through an analysis of the trajectories that take it from conception to application. I question the specific shared assumptions and beliefs that need to be assimilated between designers and users for self-tracking systems to be functionally and ideologically coherent. Here design is thought of as a verb rather than a noun (Miettinen, 2008, p. 219), it is not simply a matter of good or bad design but rather it is something that is *done* and done with *intention.*

As a practice, design absorbs ideology through discourse (van Dijk, 2006, p. 133) and as a systemic process with inputs, outputs and resources that guide it, it uses discourse to access and translate ideologies. As a research domain, design reflects a meta-ideology resulting from the collection of different discourses that are brought together and channelled towards production processes that sees them embodied in the form of artifacts and at the level of somatic bodies, data-bodies and the body politic. The discourse-analytic approach taken in this book is not one that merely seeks to uncover the underlying ideologies informing the design of self-tracking technology, but rather one that hopes to illuminate techno-cultural hegemony and the manner in which it is reproduced (Lupton, 1992, p. 149) in the evolution of self-tracking systems. Given the contingent nature of self-tracking system design – where there are dependencies on commercial actors for components, software, storage, hosting and so on – and many intermediaries involved in its implementation, ideology is perpetuated by mechanisms that are not entirely transparent. These mechanisms can be hidden from designers by way of the briefing process or they can be hidden from other stakeholders by the normal conventions of design processes and contractual arrangements between clients and vendors.

Discourse shapes how designers, users and researchers engage with self-tracking in terms of: what it is that they do, what activities constitute good design, proper usage and adequately critical research, the norms and values that are upheld in each of these three domains, and the resources that one would expect to have access – or privileged access – to in order to undertake work in each. The discourses surrounding practices in each of these domains reflects a 'web of activities that surround objects rather than the things themselves' (Prior, 2003, p. 2). The 'thing-ness' of technologies, bodies and data, and the matter of involvement with the conditions that bring them into being as situated products designed and used collectively in dynamic social settings. In choosing to analyse discourses about self-tracking including social media posts, blogs and blog comments, patents and others, I have considered questions such as who has authored and contributed to perpetuating certain ideas about the quantified self and how they are, and can be interpreted (Prior, 2003, p. 26) *against the grain* re-framing certain narratives as alternative views that help us see the forces that shape the social and emotional dimensions of self-tracking practices.

Annemarie Mol reminds us that '*ontology* is not given in the order of things, but that, instead, *ontologies* are brought into being, sustained or allowed to wither away in common day-to-day sociomaterial practices' that constitute patterns of technology use (Mol, 2002, p. 6, their emphasis). And in order to shed light on these practices, I draw on Donna Harraway's idea that:

> …an 'ethnographic attitude' can be adopted within any kind of inquiry, including textual analysis. Not limited to a specific discipline, an ethnographic attitude is a mode of practical and theoretical attention, a way of remaining mindful and accountable. (Haraway, 1997, p. 39)

The notion of a 'quantised self' that I wish to advance in this book is both a metaphor and a logic for the technocratic governance of human bodies and lives. I want to convince the reader that is something that we should acknowledge and treat with intellectual curiosity, theoretical openness and trans-disciplinarity when we do research into self-tracking and with precaution when systems are being designed. It is something that should be accounted for in technology design by viewing it as an outcome *to design against*. The 'quantised self' extends the socio-material outcomes of mediation such as alignment, error correction smoothing out and the removal of imperfection, which are effects that are commonly observed when varying phenomena such as light, sound and shapes are rendered in digital form. I borrow this term from audio production technology (e.g. drum machines and MIDI sequencers, see Lanier, 2010, pp. 9–10) where the quantise feature is used to correct human variation and snap drum hits into precise and predetermined grids. In image processing quantisation is the lossy file compression that occurs when a range of values are flattened into one such as when the number of colours in an image are flattened into a few or when brightness variation values are rounded the resulting image is an inexact approximation that is made from partial data. These techniques are irreversible and their effects are not easily perceptible to the observer.

In summary, the decision to study the politics and possibilities of self-tracking through ideologies that shape design has been undertaken here to complement, nuance and extend existing studies of self-tracking. Its outcome, if read in the spirit that it is written is to see the outlines of a mutual praxis of critique and design that contributes to facilitating the (re)imagination of political effects (and affects) that are embedded into quantified, quantifiable and quantised selves as they are fashioned and perturbed by the data-driven logics that are governing everyday life. It seeks to answer three questions:

(1) What is the quantified self, how can it be situated as a data capture regime in space and time, and how has it become an ideological frame for the data-driven governance of bodies and lives?
(2) Through what sociotechnical trajectories are self-tracking technologies finding their way into health, home and work and refashioning the way that bodies are defined and represented, and, how can tracing these trajectories expose self-tracking as inherently political in terms of whose interests it serves?
(3) How can understandings of power relations embedded in self-tracking be developed in ways that might be assimilated into the existing and commonly used methodologies to design sociotechnical systems?

The book unfolds in three parts, each emphasising one of the questions above: (Part 1) *Visions: Quantified and Quantifiable Selves*, (Part 2) *Rules/Affordances: The Embodied Practices of Self-Tracking* and (Part 3) *The Quantised Self: Discipline and Design. Visions* is about the ideology of datafication and its relationship to technology and self-experimentation. It looks at how the emergence of the 'Quantified Self' a cultural movement founded on these ideals about a decade ago, came to intersect with and then encompass and intersect with a wider trajectory of technology use for the purposes of self-surveillance, health vigilance and self-responsibilisation. Borne out of the notion of the quantified self is a form of personal data capture that has been picked up by healthcare institutions under the auspices of personal, participatory and precision medicine. The various divisions of labour and representations of bodies are examined through Nicholas Jewson's notion of 'disappearance' of the body in order to set the stage for the discussion of self-tracking in other contexts beyond healthcare.

In *Rules/Affordances* recent developments in self-tracking are placed in a broader historical context. First, it looks at the evolution of electronic monitoring for the purposes of house arrest and home confinement as a key antecedent to contemporary applications of self-tracking technology. It argues that there are important lessons to be learnt from it as a case study. It then turns to the use of self-tracking to measure social interaction in the workplace. It argues that as an extension of practices that one comes to expect at work – such as being audited, having to record and account for time spent on tasks or having communications monitored – the use of self-tracking technology blurs the boundaries between work/life, data/body and self/other in ways that make it difficult to for persons to act with agency.

the analysis and design of self-tracking systems (see Pigrum, 2012). Moreover, it is namely, his account of a dialectical relation between qualitative and quantitative representation and the competition between different types of rhythms as a source of alienation-by-alignment – what I refer to as *the quantised self*. These ideas can inform the way we think about the designer-user relation in the develop of socio-technical systems that afford self-tracking and can be adapted and harnessed to encompass a broader range of subject positions and spaces (see Apperely, 2011) It attempts to form linkages between human physiology, behaviour and the environment in a way that places an emphasis on connection between past, present and future selves and emphasises medial properties of the body which can co-exist in multiple and overlapping configurations within the logic of sociotechnical systems. Using examples of distraction and posture management technology, the chapter shows that like rhythmanalysis, self-tracking is a method of human-centric analysis that conceives of bodies, places, times and objects that are rhythmically entangled.

In the final chapter, I consider design as something that is engaged in as a process that has within it enabling and limiting forces that impact the agency of designers influence decisions that translates to the lived experience of self-tracking technology. It argues that an intervention into the outcomes of self-tracking technology requires an examination of the different perspectives on design that shape the way we might think about the positivist ideologies of self-tracking as forces that shape understandings of patterns and outcomes of use and how they are translated as priorities in processes of design and production. As a coda to the book, the chapter proposes conceptual tools that provoke reaction on the idea that self-tracking technologies in their own right can raise questions about the purposes of their existence and the ends that they achieve, and further, if speculative thinking about utopian and dystopian futures of self-tracking can be incorporated into the broad set of approaches understood as *design thinking* (see Buchanan, 1992). The chapter critiques perspectives on design processes in terms of their foci, what they achieve and where we might situate the loci of power that design practices ultimately serve. It links this to modes of data collection that are available as the basis of self-tracking system design and attempts to reframe these concepts through a dialectics of instrumentality (see Khatchatourov, Chardel, Peries, & Feenberg, 2019).

Part 1
Visions: Quantified and Quantifiable Selves

Chapter 1

The Quantified Self

> There's a lot more to self-tracking than just the Quantified Self; these days there's a lot more to 'quantified self' (lowercase) than just the Quantified Self (title case) […] the Quantified Self came to exist because people were *already* self-tracking and some of those people were interested in discussing their self-tracking experiences with others. While the Quantified Self does undoubtedly help spread interest in self-tracking (just as increasing interest in self-tracking helps drive the growth of the Quantified Self), I think the group's more significant cultural impact has been to make the very concept of self-tracking more visible […] and in doing so, making tracking-in-general more visible. It is this last function, of making more visible a particularly disconcerting thing that usually fades into the background (e.g. being tracked by others), that I think is at the heart of some people's deep discomfort when I say, 'Quantified Self' out loud. (Boesel, 2013)

> There is much still to be explored in terms of social and cultural analysis of the Quantified Self and its related movements. How do discourses and practices of quantified self represent the self and the body? What assumptions beliefs and values are drawn upon to portray the Quantified Self movement's own descriptions and commentaries in popular culture more generally. (Lupton, 2013, p. 26)

As a means of initiating a cultural analysis of self-tracking, this chapter seeks to disambiguate the Quantified Self (title case, cultural movement) from the quantified self as a more general label for a range of self-tracking practices such as life-logging, auto surveillance, biohacking and those who engage in them. The chapter's focus is on the ideologies and discourses of the Quantified Self and it aims to help the reader understand and situate the Quantified Self, its discourse and its practices within the context of the argument that is developed in the chapters that follow.

The Quantified Self is described by one of its co-founders, *Wired's*, Gary Wolf on the business-oriented social networking site, Linked In as a:

> California based social enterprise that helps people make personal discoveries using everyday science. [That works] with

The Politics and Possibilities of Self-Tracking Technology, 13–33

Published under exclusive license by Emerald Publishing Limited
doi:10.1108/978-1-80043-338-020211001

> an international community of users and makers of self-tracking tools, [...] who specialise in helping people advance their projects [and provid] appropriate technology, methods and peer support. (Wolf, n.d.)

I want to begin this chapter with an example of one of the personal discoveries that Wolf is referring to in his description of the Quantified Self quoted above. It is personal account of self-tracking shared by Whitney Erin Boesel, a humanities and social sciences scholar with an interest in self-tracking. Boesel has also been a Quantified Self participant over the years and written about these experiences in both academic and non academic forums. At the Quantified Self conference held in Amsterdam in 2017, Boesel shared a story about how she became interested in self-tracking. In her talk, *My Numbers Sucked but I Made This Baby Anyway* she tells the audience that after suffering a concussion in 2014, post-concussion syndrome left her with some symptoms that her doctors were unable to explain. Her story is both typical of Quantified Self activity but also unusually reflexive. She recounted:

> ...for a number of reasons, I suspected something hormonal, and so I started turning my capillary blood, saliva, hair and even blood serum into data hoping that it might be able to find an answer. I looked at my pituitary hormones, sex hormones, and adrenal hormones in addition to a number of other substances. One thing that I never intended tracking, however, was my Anti-Mullerian Hormone (AMH). I'd never even heard of AMH until August 2015, when I found it at the bottom of a lab requisition by one of my doctors. I looked up AMH and learned that in an adult woman it comes from [her] eggs [...] the AMH test is ordered for two main reasons [...] because it can be an indicator of Polycystic Ovarian Syndrome [or because] fertility specialists order it looking for low AMH because some clinics won't accept patients with an AMH below 1.1 because those patients don't respond well to the medication needed to harvest eggs. (Boesel, 2017)

At the time, Boesel was 35 years old. Her doctor had ordered a fertility test. The results showed that her AMH was low for her age and that based on the numbers it was likely that it was too late for her to have children. After doing some research into AMH of her own, Boesel started to develop a hunch that her low AMH might have had something to do with Vitamin D. But, after taking some supplements for a while her AMH got worse, not better. She tried acupuncture and that didn't help either. She continues:

> I was now 36 years old and I wasn't even within the reference range for women aged 36 to 40. I felt like I was helplessly watching the door shut forever on something that I [had] come to realise was actually critically important to me. As I talked through the

> numbers with my acupuncturist, however, I realised that what I was searching for in all my hormone data was not the possibility of becoming pregnant, but the permission to become pregnant [...] What I was searching for in all of my hormone data was either permission to keep putting off making a really difficult decision, or permission to seek extra help in becoming a parent despite the really bad timing. But these were social questions and in the end, no blood test was going to answer them. (Boesel, 2017)

Wolf sees a great deal of potential in stories of transformation like this one. For him they are perfect demonstrations of the full potential of the Quantified Self's idea of a 'better life through numbers'. But, Boesel's story is unusually self-reflexive and isn't one that has the technological determinism – a view which sees technology as the key driving force in the self-knowledge that is obtained through quantification and tracking – that often accompanies many of the narratives shared at the local, open and informal QS Meetups. At these Meetups, participants are encouraged to report on experiments they have conducted on themselves using data collection and analysis techniques in order to improve some aspect of their lives. The structure of this shared discourse is guided by three prime questions: *What did you do? How did you do it? What did you learn?*

For example, Ka-Ping Yee a software engineer who worked for Google tracked most of his daily activity for three years during his time in graduate school in the mid-2000s. He did this by using a widget that he wrote. In a text box situated at the upper right-hand side of his computer screen he was able to enter a brief log of his daily activity by keying in a few words and a time stamp was automatically added to his entry. The duration of his daily activities is calculated by subtracting the time from when one activity began from the time a subsequent activity commenced. The widget allowed Yee to edit text which meant that he could note activity that had taken place while he was away from his computer, and by using a standardised set of keywords in the descriptions of his activity he can easily run a script to graph the allocation of his time on any given day.

We know this about Yee because in September of 2008 he shared some insights about the ratio of time he spent writing versus sleeping while completing his dissertation at a Quantified Self (QS) meet-up in the San Francisco Bay Area and a report about his talk was written up and posted to a website maintained by the organisers of the group that hosted this event (Wolf, 2008). Five years later, in August of 2013 at a QS meet-up in Boston, Matthew Ames described to the audience his self-tracking project which had dramatically changed his weight and fitness levels (Ramirez, 2013). Unlike Yee, who had built his own software, Ames had used a digital body composition scale in combination with a Garmin Forerunner smartwatch and number of devices and smartphone applications including: Weight Watchers; Run keeper; MyFitnessPal, and the Nike + system to support his physical transformation.

These brief examples show us that the meanings derived from QS discourses are situated with respect to the particular set of activities, attitudes towards technology, specialised expertise and social relations in which this type of self-tracking activity occurs. Such discourses when examined retrospectively do not necessarily need to be taken as an expression of a participant's true beliefs, attitudes and behaviours or as free of their context and ahistorical. Rather they are the discourses of a given social context and they reflect certain ideas about technology, society and everyday life *which are in and of their time.* And in light of this situating discourse in a socio-historical context matters in the analysis of cultural movements and their lasting impact on industries and societies.

Stories like the ones described here can be drawn on to characterise the relationship of the QS′ leadership, community, and the traces of their activity to the broader 'field' (Couldry, 2004) of self-tracking by taking a 'praxis logical' (Renzi, 2008, p. 74) approach. Taking this approach allows for an alternative understanding of the QS and related movements, communities and initiatives to be developed through critiques of the functions served by activity that goes into the articulation of its ideology and guiding principles. This requires, according to Renzi (2008), attempting to address the questions: Who has put the concept forward? What are the guiding ideologies and how are its discourses organised? Where and how has the QS come into being? Under what circumstances has it come into existence and gained attention? How do the QS′ ideas circulate within its network? And how are they portable to contemporary culture as a proxy for all types of tracking? In what follows, I apply a praxis logical line of inquiry to the QS. First, I look at its birth and growth, I then turn back to examine its roots in Californian counterculture and end the chapter with a discussion of more recent QS activity.

The Birth and Growth of the Quantified Self

People get involved in the Quantified Self for all sorts of reasons. They may suffer from chronic illness and want to explore different options for treating and managing their symptoms. They might want to understand and change certain patterns of behaviour. They might be enthusiastic about new technologies and data analysis techniques. Or they might approach the QS with an entrepreneurial mindset and want to draw from the community to develop, refine or seek capital for new products and services.

At a Quantified Self Meetup, one is likely to be exposed to a range of techniques in self-experimentation and data analysis with applications in behaviour monitoring, life-logging, location-tracking, digitising bodily information, sharing health data, attempts at medical self-diagnosis, personal genetic sequencing, measuring chemical body load counts, psychological self-assessment, tracking mood and social interactions, sleep studies, measuring productivity and a number of others.

The first documented QS meeting was held in September of 2007. It was hosted by Kevin Kelly, the founding editor and 'senior maverick' of *Wired.* The meeting, was a gathering of 28 people at his home in the San Francisco Bay Area. The group assembled was comprised of people almost exclusively from the technology industry. Kelly had brought the group together to discuss the application of objective measures of human physiology and behaviour for the purposes of self-discovery and management. Among those gathered were Gary Wolf, Seth Roberts, a psychologist who had gained notoriety for experiments he had conducted on himself to try to improve issues with his sleep, mood and weight and Tim Feriss, a technology entrepreneur and author of a popular self-help book, *The Four-Hour Work Week.*[1] At this meeting, Kelly, told the audience:

> In 2007 we began looking at some new practices that seemed, loosely to belong together: personal genomics, location tracking, biometrics. These new tools were being developed for many different reasons, but all of them had something in common: they added a computational dimension to everyday existence. Some of this was coming from the outside as marketers and planners tried to find new ways to understand and influence us. But some of it was coming from inside as our friends and acquaintances tried to learn new things about themselves. We saw a parallel to the way computers, originally developed to serve military and corporate requirements, became a tool of communication. Could something similar happen with personal data? (Quantified Self, 2012)

[1]Self-help books are predicated on the assumption that self-directed reflexive analysis is required to be self-efficacious and to achieve the book's ultimate aim of developing a personal philosophy or approach to living that aligns to the objectives of the book. These might be the accumulation of personal wealth, being healthier or fitter, overcoming addictions, forming more enriching relationships or becoming more successful at work for example but there are self-help books that cater to a wide range of audiences and range from being broadly focused or very niche. Inherent in this logic is also the idea that to become self-efficacious one must also confront 'home truths' about themselves and their lives which when the reader recognises enables the reader to objectively view some or another personal trait that is standing in the way of what might be considered as a healthy, productive, resilient, or responsible way to live. The invitation in many of these books to reflexively analyse one's life by way of techniques of measurement, journaling and inventorying in order to track progress towards goals and be constantly thinking ahead about how one's habits and behaviours in the present may have longer-term effects (Rainwater, cited in Giddens, 1991, p. 2; Burrows & Nettleton, 2002). Reading self-help books are an individual self-directed undertaking but it also involves interpellation of the reader where peer support is displaced into a broader network of relations that occur within or outside of the auspices of group therapy or mutual aid and simply places this into the realm of people who have read or are reading the book. Self-help and self-tracking are practices that invite one to confront a reality that is expressed in terms where one sees themselves in information being presented to them in the form of a narrative, metaphor, statistics or data. This means that the individual's motivations, behaviours are guided by something more concrete than faith or intuition in data-driven forms of self-help.

Writing in *Wired*, Wolf has also reflected on this meeting and on the origins of the group:

> Two years ago, Kevin Kelly and I noticed that many of our acquaintances were beginning to do this terrible thing to themselves, finding a clever way to extract streams of numbers from ordinary human activities. A new culture of data was taking shape. The immediate cause of this trend was obvious. New tools had made self-tracking easier. In the past, the methods of quantitative assessment were laborious and arcane. You had to take measurement manually and record them in a log; you had to enter data into spreadsheets and perform operations using unfriendly software; you had to build graphs to tease understanding out of the number. Now, much of the data-gathering can be automated, and the record-keeping and analysis can be delegated to a host of simple web apps. (Wolf, 2009)

At a TED talk in 2010 in Cannes, Wolf told the audience that as sensor-enabled technologies were becoming more readily available to everyday people, there was going to be great potential when these sensors would get turned inwards and used to help people become more enhanced, self-knowing, beings. A significant portion of this talk deserves quotation here as Wolf's Cannes talk (2010a), along with a *New York Times* article titled *The Data Driven Life* (2010b) have been instrumental in establishing a powerful frame of reference which repositioned self-tracking as a self-directed, agency-enabling practice. Wolf opens with the following parable:

> I got up this morning at 6:10a.m. after going to sleep at 12:45a.m. I was awakened once during the night. My heart rate was 61 beats per minute, my blood pressure, 127 over 74. I had zero minutes of exercise yesterday, so my maximum heart rate during exercise wasn't calculated. I had about 600 milligrams of caffeine, zero of alcohol. And my score on the Narcissism Personality Index, or the NPI-16, is a reassuring 0.31. (Wolf, 2010a)

Wolf smiles, a few people laugh, but the overall audience response could be read as lukewarm. Wolf goes on to explain to the audience that:

> A few years ago, Kevin Kelly, my partner, and I noticed that people were subjecting themselves to regimes of quantitative measurement and self-tracking that went far beyond the ordinary, familiar habits such as stepping on a scale everyday. People were tracking their food via Twitter, their kids diapers on their iPhone. They were marking detailed journals of their spending, their mood, their symptoms, their treatments. (Wolf, 2010a)

Wolf informs the audience that the improvement and reduction in cost of biometric sensors such as 3D accelerometers that can track movement through space meant that they could be integrated into all sorts of digital devices and be used to generate incredibly detailed information on things like physical activity and sleep. He shows the audience an image of one early adopter of this technology, the now market leading, Fitbit. He shows the audience a headband that is designed to track sleep. He compares this to an example of a sleep tracking system developed a few years prior noting how heavy, cumbersome and uncomfortable it would be to use. He then tells the audience that he has just presented some Quantified Self experiments at a healthcare conference in Washington D.C involving a modified asthma inhaler fitted with a GPS transceiver which records the date and location of an asthma incident, giving the user a new awareness of their vulnerability in relation to time and environmental factors that could be correlated from other data sources such as published pollen counts and pollution indexes. He continues:

> Now we know that new tools are changing our sense of self in the world. These tiny sensors that gather data in nature, the ubiquitous computing that allows data to be understood and used, and of course the social networks that allow people to collaborate and contribute. But we think of these tools as pointing outward, as windows and I'd just like to invite you to think of them as also turning inwards and becoming mirrors. So that when we think about using them to get some systematic improvement, we also think about how they can be useful for self-improvement, for self-discovery, self-awareness, self-knowledge.

Writing in the *New York Times* in the same year, Wolf (2010b) describes a range of Quantified Self experiments. Most of them involve men who work with computers trying to deal with what we might colloquially refer to as being first world problems. Wolf talks of blind spots in our lives, things slipping from our attention, the fallibility of memory and gut-feels which when ignored cause us to make errors of judgement. While Wolf notes that these men all know their behaviour is abnormal [and that] they are outliers, he also constructs these practices as forward thinking and already becoming increasingly normal thanks to sensors getting smaller, better and cheaper, the popularisation of smartphones, social media making it normal for people to share intimate details about their lives enabled by a trajectory of technical innovation that Wolf expresses as part of a rise of 'global super intelligence known as the cloud' (2010b). Wolf writes:

> I had recently written a long article about a trend among Silicon Valley types who time their days in increments as small as two minutes, and I suspected that the self-tracking explosion was simply the logical outcome of this obsession with efficiency. We use numbers when we want to tune up a car, analyse a chemical reaction, [and] predict the outcome of an election. We use numbers to optimise an assembly line. Why not use numbers on ourselves?

> But I soon realised that an emphasis on efficiency missed something important. Efficiency implies rapid progress toward a known goal. For many self-trackers, the goal is unknown. Although they may take up tracking with a specific question in mind, they continue because they believe their numbers hold secrets that they can't afford to ignore, including answers to questions they have not yet thought to ask. (Wolf, 2010b)

After a brief period of growth, the number of active participants in official Quantified Self Meetups had declined significantly though its brand, its ideology and the frame that formed around it was extending outwards into the discourses of start-ups, large corporations the healthcare sector and the lifestyle technology market. Drawing on data that I collected manually from the Quantified Self and QS Meetup sites, we can see that in its early phase between 2007 and 2010, there were Quantified Self Meetup communities in San Francisco and New York. By 2010 nine groups were established in Amsterdam; San Diego; Seattle; Boston; Chicago; Toronto; Sydney; London, and Brussels. The biggest period of growth for Quantified Self meetings was between 2012 and 2013 which is when the number of groups grew from 50 to 108 worldwide.

From his writings on the Quantified Self it is evident that Wolf is quite taken by the work of psychologist, Albert Bandura, particularly Bandura's work on the concept of self-efficacy (see Bandura, 1982). The aspect of self-efficacy theory that seems to appeal most to Wolf is Bandura's idea that the successful achievement of one's goals is something that is not entirely dependent on personality traits or the beliefs that one has about themselves, but rather it has more to do with the *expectations* that people have about how likely they might be at succeeding in making progress towards their goals and aspirations. Wolf also has his own theories on self-efficacy and these are formulated by his interpretation of some of Bandura's ideas. Wolf believes that self-efficacy can be measured, correlated to probabilities of success or failure and that these probabilities can be altered through behaviours that were persistent, tenacious and approached in an evidence-based fashion.

Wolf's interpretation of Bandurian self-efficacy operates on the premise that if self-efficacy has something to do with orienting oneself, and one's body, relative to their environment and other people. If this was the case, then one needed to have a 'generative capability in which the component cognitive, social and behavioural skills must be organised into integrated courses of action to serve innumerable purposes' (Bandura, 1982, p. 122). For Wolf when these courses of action are related to aspects of personal health and wellbeing, they can become self-regulatory forces that could facilitate the exercise of self-influence over one's 'own behaviour by means that extended beyond sheer willpower alone' (Bandura, 1982, p. 129).

The data that is collected through self-tracking can, for Wolf, can be used to support the development of the persistence, tenacity, motivation and self-discipline required to become more self-efficacious. This 'notion of self-efficacy [was] separated from the more general concept of self-esteem, [and it could be] easier to notice where low self-efficacy may interfere with learning or achievement [or lead to] avoidance behaviour' (Wolf, 2007). For Wolf, meeting personal goals in a consistent and methodical way could be sustained if people could see change occurring, or progress towards goals accumulating over time. Wolf believed that self-efficacy could be improved, and previously inaccessible potentials could be realised if people had better tools and methods to measure, see and correlate the variables that related to the goals that they had set for themselves. The issue with self-efficacy, as Wolf sees it, is that given it is a concept concerned with the realisation of individual goals, with the increasing complexity of relations between health and environmental variables, 'self-efficacy [could] get lost in the noise [created by] general self-esteem, physical tiredness, environmental stress and other distractors' (Wolf, 2007).

The solution to this problem of noise, for Wolf, is to turn to objective measures that could be supported by the development of 'instruments for measuring and tracking self-efficacy' and that these would be 'crucial for regulating it and progressing effectively towards set goals' (Wolf, 2007). Stirred by the growing enthusiasm for using spreadsheets, sensors, data and information visualisation techniques, Wolf sees great potential in self-measurement and advances in sensor technology were

> ...accelerating along a number of different dimensions: especially, the number of measurements that could be taken per unit of time, and its corollary that measurement could also get closer and touch the deepest confines everyday life. (Wolf, 2007)

If Wolf's perspective on self-tracking has evolved from an interest in the role of technology in the psychology of attaining personal goals then Kelly's interest in the Quantified Self has more to do with techno-libertarianism and technology as a means for fostering greater self-sufficiency. Kelly tells the audience at a QS Meetup in San Francisco:

> The thing about the Quantified Self is that it's two words so I just wanted to talk about the two halves, quantification and the self. The quantification [bit] is what everyone is focused on and that's where there's the most surprise. We have increasing tools, and I'm somebody who believes that science is changed by tools, when we had a microscope or a telescope that really opened up worlds that were unseen before and new tools give us new questions, but it's [still] the new tools that change things the most. And so, the Quantified Self is being built on the ability to quantify things, the technologies of quantification are allowing us to do things as part of science that we have never done before. There has always been introspection and we've known that there are many other ways to know yourself without quantification: meditation, self-reflection, journaling, art [...] but the new science of these tools

> allows us to use quantification as a means of knowing but it's not just ourselves that will change. (Kelly, 2012)

At the same meeting, Kelly offered his views on how *he* thought the Quantified Self *should* evolve:

> One of the things we're trying to do with this, and we've been very careful about structuring it, is [...] I still don't exactly know what the Quantified Self is. The whole purpose of having these monthly meetings is to have everyone tell us what it is. We're using this whole thing to define it. Each time someone comes up here and tells us about what they are doing, why they are doing it and what they learnt [...] which is what the format [of the presentations] is [...] they are helping us to define what this Quantified Self business is. So rather than having a preconceived idea [about what the Quantified Self is] we had a hunch about the direction, about a general drift, which is obviously [about] self-monitoring [and] self-tracking but as to what is and what is not [the Quantified Self] that's up to the people who show up here and give talks, and it's a self-propelled movement in that sense. So [...] we are very hands off and we have meetings all over the world. Now each one who stands up [at a meeting] is telling us a definition of what [the Quantified Self] is. (Kelly, 2012)

For Kelly, the work of the Quantified Self had the potential to challenge dominant scientific paradigms. Later in the same speech, Kelly tells the audience:

> Part of the whole excitement of the Quantified Self is that we're changing some of the ways we do science. An example of that would be in the normal way of doing science, you're dealing with many examples, you're studying populations or many cases. But in the Quantified Self you have an n-of-1 and that is a different kind of science [for example] the gentleman scientists of England and Europe. But in other ways, the science of the Quantified Self is engineering some other different methods in how we know things. The standard dogma of science is that you experiment on a single variable. You isolate one thing but if you're doing an n-of-1 experiment you don't have time, so you do experiments on multiple variables and [then] use math to unravel the fact that you have multiple variables. (Kelly, 2012)

By the end of 2011, a further 16 groups had been established, including a separate group in Silicon Valley and outside the United States, Canada and Europe, a group of 36 self-quantifiers was established in Beirut. In 2012 a further

32 groups were established, predominantly in the United States and Europe, four in Asia, one in the Middle East, and one in Australia. In 2013 22 groups were established including two in India and Latin America respectively.

In my hometown of Melbourne, Australia the first Quantified Self Meetup was held on January 30 of 2013. It was held on campus at The University of Melbourne and organised by the Interaction Design Lab which is part of the Faculty of Engineering's School of Computing and Information Systems and the Health and Bioinformatics Research Unit which forms part of the Faculty of Medicine, Dentistry and Health Sciences' Centre for Digital Transformation of Health. These are both parts of the university that are usually devoid of researchers from the Humanities and Social Sciences. A participant talks of how keeping a pain journal using a pen and paper to manage a chronic knee injury had motivated them to track productivity, mood and sleep. Another presenter talks of using Google Sheets and Wolfram Alpha Pro software to track fitness variables such as heart rate variability, step counts and sleep to improve their rock-climbing performance. A third speaker, a professor of health informatics presents 20 years of marathon and training data to demonstrate a model that allowed him to predict run times for various distances based on past performance.

By 2014 only seven new groups were established, five in 2015, three in 2016, and one in 2017 and 2018 respectively. Most of its membership numbers are made up of groups where people signed on, perhaps attended a few meetings and then lost interest. After this many groups became dormant. At the time of writing, most of the Quantified Self groups appear to be inactive. Ten groups have been inactive since 2013, 12 since 2014, 17 since 2015, 15 since 2016, 16 since 2017, and over the period between January 2017 and June 2018 twenty-six Quantified Self Meetup groups have been active, with 10 groups having held an event in 2018. Of the 110 Quantified Self Meetup groups that formed between 2007 and 2018, 19 were sustained long enough to have one or two meetings. The most active groups were London (UK), The San Francisco Bay Area, and Lansing (Michigan) with 63, 59, and 55 meetings held in the same eleven-year period. A Membership data beyond 2018 was not available at the time of writing. Eight groups grew to have over 1000 members: San Francisco (1052); Silicon Valley (1246); Berlin (1356); Amsterdam (1415); Boston (1452); London (2353); New York (2511), and the Bay Area (4644).

It is important to note that Meetup group memberships are not an accurate representation of active participants. On average, a small number of people who physically attend and actively participate in the meetups themselves. At the time of writing, of the 110 Meetup group pages 48 had been deleted and data for these groups had to be gathered using versions of their pages on the internet archive's *Way Back Machine* although these inactive sites are still listed on the Quantified Self website which to the casual observer gives the impression that the community is larger than it actually is.

In June of 2014, Apple's CEO Tim Cook announced that the latest iteration of the mobile phone operating system for the iPhone, iOS eight would include a health application and developer platform named *Health Kit*. The software had been designed to serve as a hub for various health tracking applications and wearable health tracking devices to centrally store and integrate user data. *Health Kit* would allow for applications that monitor health related variables to house data in a central location on the device that could be shared directly by the user, or with permission, to hospital software systems or directly with healthcare professionals. The health application and developer platform allowed applications that monitor indicators of health to be housed within the device for sharing with hospital software systems or directly with individual healthcare professionals. Shortly after the launch of a developer platform, the Quantified Self's leadership released an app for Apple's iOS platform that allowed users to access and export their *Health Kit* data in a tabulated format so that it could be analysed in spreadsheets, with the statistical coding language, R, or any other tool that could read a Comma Separated Values (CSV) file.

By the time the QS15 conference was underway, the Quantified Self were positioning themselves as a well-respected community of life-loggers, self-trackers, biohackers, data visualisation experts and health advocacy group. They were also seeking to become an established player in the industry for personal metrics. There were 39 commercial sponsors supporting the conference many of them featured in the talks that made up the official program. Among the sponsors were companies looking to break into the self-tracking market like the design software firm, Autodesk, the geographic information systems company, ESRI, the insurance and financial services company, Suncorp and the pharmaceutical giant, Bayer. But the sponsors also included a different kind of company, companies manufacturing self-tracking technologies and services.

Then in March of 2016, Apple expanded its suite of software development kits and launched *Research Kit and Care Kit,* open source environments where medical researchers could develop tools to capture data for research purposes and through which patients could manage their own health data by tracking symptoms, creating care plans in collaboration with the healthcare professionals looking after them using the sensor hardware within Apple devices. At around the same time, QS labs was formed by Wolf and Ernesto Ramirez as a social enterprise to support the Quantified Self community. Ernesto Ramirez, formerly a community organiser, became more deeply involved in the Quantified Self as the program director for QS Labs (2012–2016 under Wolf's directorship) while he was a doctoral student in the Centre for Wellness and Population Health Systems at the University of California, San Diego 2012, describing the work of QS labs in a post on the QS website he wrote:

> Here at QS Labs, we're dedicated to helping create and grow a culture that enables everyone to generate personal meaning from their personal data. Sharing, participation, and exploring new models of discovery are core themes we're exploring as part of our QS Access work. We'll be following this initiative as it moves from today's announcement to tomorrow's reality. (Ramirez, 2015)

QS Labs ran the global Quantified Self conferences, the QS blog and users guide. Although the QS were committed to not being overly reliant on external sponsorships to support their project, they were weary of the group splintering when local chapters began accepting commercial sponsorship. They offered the following advice to Quantified Self meeting hosts:

> Please don't involve commercial sponsors without checking with us. Now that Quantified Self has so many meetings, we've had a chance to learn a lot about the expectations associated with sponsorships. Based on this experience, we recommend that you use free/volunteer locations for hosting the meetings, ask attendees to give a small donation so that you don't have to buy refreshments with your personal funds, and avoid other commercial sponsorships so that you don't risk entangling us (that is, Quantified Self Labs) in commitments we don't know about. However, if the host of your meeting would like to give you a logo to put on the sponsor spot on the Meetup page, that's normally fine. If you have special events [that] you'd like to organise with sponsors using the Quantified Self name, please let us know before you've made any commitments, so we can work things out in advance. (Quantified Self, n.d., Meetups https://quantifiedself.com/meetups/)

As I have shown here, the Quantified Self emerged out of a loose alliance of writers, hackers, artists, experimentalists and entrepreneurs from the West Coast of the United States who contributed to defining am orthodoxy for the notion of a particular type of self-tracking. It has resulted in a fusion of the culture of the communes of San Francisco with the technology industry that sprung up around Menlo Park and Silicon Valley but there are deeper roots to this ideology which are discussed in the next section.

Links to Californian Countercultural Ideology

Traces of the Quantified Self's ideology is found in both the story and pages of the *Whole Earth Catalog*, a publication founded in the mid-1960s by the American writer and entrepreneur, Stewart Brand (b. 1938). Brand had recognised that there were about 10 million people in the US living alternative, communal lifestyles. To cater to this audience's need for information about emerging technologies that could support the alternative forms of social and economic organisation that were being experimented with in the communes and countercultures of the day. *The Catalog* was launched under the auspices of providing its readers with 'access to tools'. In an opening editorial, Brand wrote that in response to the power of government, big business, formal education and religion:

> [a] realm of intimate power is developing [...] the power of the individual to conduct his own education, find his own inspiration, shape his own environment, and share his adventure with whoever

> is interested. Tools that aid this process are sought and promoted through [*The Catalog*]. An item is listed [...] if it is deemed: (1) useful as a tool, (2) relevant to undefended education, (3) high quality or low cost, (4) not already common knowledge, (5) easily available by mail. (Brand, 1968, p. 3)

Brand had been heavily influenced by American designer and systems theorist Richard Buckminster Fuller. In the opening pages of the first edition, Brand acknowledged that Fuller's insights motivated him to launch the publication. Beyond Fuller's interpretations of systems thinking, craft, community and learning, it was his extensive practice of personal archival that are a significant antecedent to the logic of self-quantification. Between 1917 and 1983, Fuller broke down his days into fifteen-minute intervals and documented his daily activities and ideas. He meticulously took notes, made sketches and collected clippings in scrapbooks. His archive, *The Dymaxion Chronofile* amounted to 140,000 pages and 1,700 hours of audio-visual material.

Fuller's intention was that the *Chronofile* be used by *others* discovering his work in the future so that they would have a better understanding of the evolution of his philosophy and the times in which he lived rather than tool for his own purposes. Fuller believed that a personal archive of this magnitude would give a much deeper insight into his philosophy than what might be gleaned from simply reading his books. Fuller's approach to personal archival was to favour completeness over selectivity or curation. This way, someone using the archive could work through it unhindered by rigid categorisations and information taxonomies and produce a variety of interpretations of his everyday life and the evolution of his thinking.

The Catalog's first edition included a number of reviews and commentaries about the various ways in which people might use technology to support sustainable alternative living, cultivating their own mushrooms, beekeeping, building tipis. Many of the books that were featured focused on Do-It-Yourself (DIY) practices and their relationship to various types of tools that could help make this happen. For example, the Volunteers for International Technical Assistance's (VITA) *Village Technology Handbooks Volume 1 & 2* are recommended as an authoritative source of 'specific practical information on small group technology' (1968, p. 18). The handbooks list various items ranging from spinning and weaving equipment, incubators, scales, flashlight projectors, photo enlargers, solar powered food drying equipment along with various types of agricultural and construction equipment. *The Way Things Work: An Illustrated Encyclopaedia of Technology* (1967) was featured for its extensive range of technologies with illustrated explanations of how they work. The book explained complex technical processes such as temperature measurement, navigation and data processing and would have appealed particular to those interested in DIY practices and appropriate technology. Industrial designer, Henry Dreyfuss' *The Measure of Man: Human Factors in Design* (1959) written for those designing objects for human use and its purpose to provide a reference point for figuring out appropriate lengths and angles of standing and sitting men, women and children. The book also contained extensive charts and tables about display and control shapes, environmental tolerance zones, reaction times, and growth statistics for the three major body types – endomorph, ectomorph and mesomorph.

These examples give a sense of how the topics of self-sufficiency, understanding how technical things and processes work and designing technology with the its users in mind were being used to educate people about how empowering and creative uses of technology could contribute to the development of alternative forms of knowledge, political empowerment and less reliance on the capitalist-industrialist system. These ideas were also being linked with others about cybernetics particularly through the ideas put forward by Norbert Wiener in *The Human Use of Human Beings* (1954) and its focus on thinkers, designers and scientists as key actors in understanding future ecological, industrial, technological and social challenges. This is a form of technological positivism that is also seen later in the early computer hacking culture of the 1980s especially in terms of its emphasis on open-source production, open hardware architectures and the non-proprietary use of information and resources.

When Steven Levy's book, *Hackers* was published in 1984 it inspired Brand and Kelly to organise the first 'Hackers' conference, a meeting of 'digital explorers from those who tamed multimillion dollar machines in the 1950s to contemporary young wizards who masters computers in their suburban bedroom[s]' (Levy, 1984, p. 4; also Brand & Crandall, 1988). The following year, Brand, Kelly and the epidemiologist, Larry Brilliant launched a spin-off publication dedicated to the internet, *The Whole Earth 'Lectronic Link* (*The WELL*). *The WELL* was an online community where members could discuss new technology (Armstrong & Hagel, 2000). *The WELL* differed from many of the online communities that were around at this time in that it did not allow users to make anonymous posts and maintained a 'you own your own words' policy that encouraged bonds to form between members and information to be shared through personal narratives and life experiences. By the late 1980s, Brand had asked Kelly to edit and oversee the production of *Signal*, a Whole Earth branded publication dedicated to personal communication tools, which at the time included fax machines, satellite televisions, mobile phones, digital retouching software, computer games and other technologies featuring in the emerging digital culture. Brand, states in a foreword:

> Kevin and the crew wish to consider the gamut of explosively diverging information frontiers as one meta-domain, intricately and subversively interconnected. Any tool or skill from one of them may inform or transfigure another one. With over half the American workforce now managing information for a living, any apparent drone drudging away on mainstream information chores might be recruited by some handy outlaw technique or tool into the holy disorder of hackerdom. (Brand, 1988, p. 3)

By the end of the 1980s, Brand had hired Kevin Kelly to edit another spin off publication, the *Whole Earth Review* (formerly *Coevolution Quarterly*). It was a journal dedicated to reporting on unorthodox and emerging trends in technology before they were picked up in the mainstream media. One of these trends was the maker culture that was forming around groups such as Menlo Park's HomeBrew Computer Club.

Gary Wolf (b. 1962) is 10 years junior to Kevin Kelly. He began working at *Wired* in 1994, a year after it had been established by Louis Rossetto and Jane Metcalfe in San Francisco. The couple had move there from Amsterdam where they had been publishing *Electric Word* magazine, an influential periodical that reported on advanced linguistic technology and global computer culture. Although it ran bimonthly editions between 1987 and 1990. The magazine would routinely featured many prominent figures in the Californian computer culture including Kelly and MIT Media Lab founder, Nicholas Negroponte. The final edition of *Electric Word* featured a photoshopped image of the American architect and designer, Richard Saul Wurman who would eventually go on to establish the Technology Entertainment and Design (TED) conferences.

Wired was financially backed by Negroponte, and for Wolf it was the main outlet for his writings on a range of issues relating to technology and society. In the first years of his time at *Wired*, this included topics such as online music repositories, and information activism. Kelly and Wolf wrote their first piece together in early 1997, 10 years before starting the Quantified Self (*Wired*, 5.03) establishing a professional partnership that eventually led to the formation of the Quantified Self a decade later.

QS 2020

In June of 2020 Wolf and de Groot had a paper entitled *A Conceptual Framework for Personal Science* published in the journal *Frontiers in Computer Science*. The paper set out to describe the different self-experimental practices seen in the Quantified Self community. Wolf and de Groot state in a news update on the Quantified Self website that this was needed because their work with academic and biomedical collaborators 'often involved [the] laborious effort' to work through misunderstandings caused by the thicket of 'somehow related terms and labels' (Wolf & De Groot, 2020) such as self-tracking, self-experimentation, n-of-1, single subject research, peer-to-peer health and the quantified self.

The aim of the article was for the Quantified Self to establish a formal presence in peer reviewed discourse that could be leveraged to initiate new partnerships and initiatives. The paper presents a five-part conceptual framework which is intended to 'guide research and education into the practice of personal science' (Wolf & De Groot, 2020). As Nils Heyen has recently written:

> From the perspective of science (and technology) studies and public understanding of science, these self-tracking research practices are an interesting empirical phenomenon that deserves further scrutiny. It seems that, far away from institutional science, some lay people or citizens, at least no professional scientists, use methods and

> procedures known from science such as research design, data collection or data analysis in order to produce knowledge for self-use in their daily lives. How can the relationship between science and society, or science and the public be characterised here? (Heyen, 2020)

The idea of personal science is deeply embedded in the QS ideology. This may have something to do with the involvement of Seth Roberts with the group in its early days and the influence that his self-experimental practices to improve his sleep, mood and weight had over Wolf. Roberts was a prolific blogger who wrote on a number of topics relating to Do-It-Yourself (DIY) health, microbiomes, the relationship between health and environment, the fallibility of experts and academic fraud so it is not surprising that Wolf was interested in his work as were many in the QS community. Roberts also respected Wolf and what the QS were trying to achieve. Roberts has argued that although the availability of new equipment can lead to new scientific ideas, his own use of personal computers to collect and store data might not fully explain the success of his experiments because laboratory scientists were also using the same tools and techniques (2010, p. 482). Roberts viewed the Quantified Self's work as a pathway to greater, more enhanced forms of individual literacy about health and the efficacy of therapeutic intervention either by healthcare professionals or by an individual on their own. Writing on the QS forum, Roberts puts is as such:

> Right now, as far as I can tell, most people feel helpless. They do what doctors or other experts tell them to do, even if it doesn't work very well. Long ago, hardly anyone could read. This left them in the grip of those who could. But eventually came mass literacy, when the benefits of reading finally exceeded the costs (because more books were available at lower prices). Reading is primitive science: if you read about things that happened, it is information gathering. It resembles doing a survey. Nowadays, almost everyone (in rich countries) reads, but almost no one does [their own] experimental science. This leaves them in the grip of those who can do experimental science (e.g. drug companies). I think my work [and the work of the QS community] suggest[s] we are close to another turning point, where, for non-scientists, the benefits of doing experiments outweighs the costs – thanks to Gary Wolf.

When Roberts died suddenly at the age of 60 after suffering a heart attack while hiking in California in April of 2014, Wolf eulogised his friend in a post on the Quantified Self forum. He wrote:

> My friend, Seth Roberts, a pioneering self-experimenter and personal scientist, died [...] the news was unexpected and very sad. A few things [that] Seth taught me: Doing lots of experiments keeps you supplied with new ideas; with sensitive and reliable

> measurements, tiny experimental effects can yield surprising clues; the people who care [most] about the problem have the best chance of solving it, if they have access to tools. Seth's contributions as a colleague and teacher had many dimensions but thinking about him nonstop this morning what I find myself marvelling at the most is the unusual style he had in nearly every conversation. Seth became interested when he saw someone thinking independently and, like the best teachers, he wanted to understand the process by which students and collaborators developed confidence in their conjectures. Countless times, I heard Seth ask somebody: 'Why did you think that?' His challenge was direct and generous, for if you were willing to expose your reasons you could count on him to apply himself alongside you, thinking up ways to improve your investigation, make your measurements more practical, or give your analysis more mathematical and logical power. Seth was acutely aware that confidence in experimental results requires investment. More than any other experimentalist I've known, Seth respected the incremental nature of building confidence in our ideas and finding new ideas worthy of confidence. He encouraged us to do small experiments first, to find easier ways to collect data, to avoid being tripped up at the outset by grandiose schemes of irrefutability.

Wolf and De Groot (2020) stress that 'personal science specifically addresses personal questions' which concern private experiences and it is the self-reflexive quality of these questions that makes the data so valuable. For example, Epidemiologist and experimentalist, Mark Drangsholt often spoke at Quantified Self events about his foray into personal science. His talks are centred around a 'health event' that he experienced during a research conference when for no apparent reason his heart rate went 'completely haywire' (Wolf, 2012) and he became unconscious. Drangsholt is a dedicated self-quantifier and had been tracking a number of variables so he was able to draw on his epidemiological tracing and his dataset, he was able analyse his heart rate variability (HRV). Using a 'case crossover' experimental design (see Kenward & Jones, 2014) which is a method that focuses on what is happening in the body just before the onset of symptoms and attempts to draw correlations between the onset of symptoms and potential environmental, chemical, physical and behavioural triggers and ascribes causality depending on whether symptoms have occurred within a reasonable time interval between exposure and onset. Using this method, Drangsholt discovered that his health events were linked to times when he had engaged in high intensity exercise, consumed caffeine, had inadequate sleep, or was about to speak in public.

Five elements are common to personal science according to the model that Wolf and De Groot (2020) put forward. The model is summarised and developed in the Table 1.1.

Table 1.1. Personal Science Model for Using Data Driven Empirical Methods in Everyday Life.

Mode	Method	Learning
Questioning – Personal questions about private life experiences and emotions.	Deliberate choice of questions based on personal objectives and methods dictated by levels of technical and data literacy and access to resources.	Self-reflexive mode of analysis where the subject's everyday life is a key part of the research domain.
Designing – Empirical approaches vary in formality and complexity.	Moving from simple self-reflection on patterns of change to complex experiments with alternating adjustments across multiple combinations of variables.	Search for cause and effect relationships and change over time.
Observing	Data can be recorded by pen and paper or digital means, can be active or passive and garnered through specialised tools built for this purpose or by off the shelf products.	Object of analysis is data from self-tracking, knowledge accumulates by structuring observations about one's own life through correlations of physiological, behavioural and environmental data.
Reasoning	Results of personal science are outputted as visualisations, graphs and tables, and also involves common formats, conventions and file types.	Interpretative processes are carried out by the individual but are conditional on the logic of tools, formulas and platforms used to analyse data. These mechanisms of interpretation may be open or opaque.
Discovering	Knowledge accumulates throughout the process data collection, analysis and reflection.	Insights gained throughout this process can lead to further questioning or abandoning of a particular experimental technique for another improved or better suited one.

Source: Adapted from Wolf & De Groot, 2020.

While the motivations to collect data and engage in self-experimental practices are both by driven curiosity and eudaimonia, and indeed a blurring of the two where scientific method could be in itself a pleasurable form of creative expression, the ancillary benefits include deeper learning about health and the generation of new ideas (Wolf & De Groot, 2020) about care, management and treatment. This brings with it questions about the role of the *individual* in n-of-1 research and the *person* in personal science. Further it raises questions about the politics of information in personal science and data-driven approaches to healthcare as data is transacted across social spheres, technical infrastructures and the expert-lay divide.

We know that Quantified Self participants are already highly 'knowledgeable in ways that could fruitfully inform' (Irwin & Michael, 2003, p. 111) clinical and policy level work in the healthcare sector. It is the blurring of knowledge which sees individuals who self-track become embedded – both as real persons and as they are represented by data – and projected into networks where alliances between the technology industry and institutions immerse individuals into mechanisms of value extraction that are not entirely declared or transparent. Those engaging in self-tracking are at once embodying and performing physical, epistemological, pedagogical, immaterial and interpretative forms of work whilst drawing on a convenient established culture of technology where there is a blurring of creative expression and experimentation and embodied performance within tightly scripted sociotechnical systems.

As personal science extends deeper into the mechanism of knowledge production that dominates contemporary market-driven, neoliberal healthcare, admixtures of technical, ethical, scientific, regulatory and design discourse will be drawn on to build cases for the extension of self-tracking deeper into areas of social life. This brings with it a dynamic where individuals are engaged in the co-construction of hybrid data selfhood and subjectivities over which they are likely to have little control or at least forms of control that are conditional on and governed by technical and technocratic logics that form by way of the imaginations of those sponsoring self-tracking systems and those that engage in its advocacy and development.

It is easy to take aim at those who self-track as upper middle class, urban-centric, careerist people who form part of a 'self-absorbed technical elite who use arsenals of gadgets as a kind of self-imposed panopticon, generating data for data's sake' (Nafus & Sherman, 2014). By virtue its data-driven, self-experimental and personal-scientific aspects, self-tracking practices afford participants their efforts from other forms of self-betterment that may be more fairly labelled as vanity and self-obsession. Yet, still the outcome of self-efficacy and goal realisation rests on externally set standards and norms. As one approaches a point where it appears as if they are reaching their goals they will have arrived at this point through an accumulation of feelings of inadequacy and anxiety (Duval & Wicklund, 1972) which are not likely to simply disappear once someone feels like they have made progress. Wolf's The Data Driven Life article in the *New York Times* attracted about 140 comments after moderation. The debate that unfolded between April 28 and 7 May 2010 is an interesting one. Especially with a decade of hindsight. For instance, one commenter asks:

> Who is 'we', Mr. Wolf? 'We are in the process of changing'? So … is there a 'they'? (And what would you call them, Luddites? Technophobes? The poor? Those on the other side of a bridgeable digital divide?) There are other views of 'we' […] This is only to say that 'we' is a problematic assumption. Also, it might be good to look into the pros/cons of 'ubiquitous computing' to get a little more critical distance.

It is hard not think about how well suited the Quantified Self's their methods of n-of-1 experimentation are to a 'calculating frame of mind' (Featherstone, 1982, p. 26 cited in Turner, 1984, p. 168) and how this frame of mind a supports the objectives of neoliberal health by encouraging the capture of data from large groups of people and the sharing of this data with third parties for the purposes of extracting value. And although Wolf himself acknowledges that:

> …self-tracking can sometimes appear narcissistic, but it also allows people to connect with one another in new ways. We leave traces of ourselves with our numbers, like insects putting down a trail of pheromones, and in times of crisis, these signals can lead us to others who share our concerns and care enough to help. (Wolf, 2010b)

There is also an effect associated with this way of thinking about self-tracking which upholds the objectives of neoliberal health and its automation of set of processes that are involved in the production of clinically relevant information. At the level of a more macroeconomic doctrine, the concern about innovative cultural movements and their potential to be genuinely counter hegemonic is reflected in something that Raymond Williams said over four decades ago, which is:

> […] however dominant a social system may be, the very meaning of its domination involves a limitation or selection of the activities it covers, so that by definition it cannot exhaust all social experience, which therefore always potentially contains space for alternative acts and alternative intentions which are not yet articulated as a social institution or even project. (Williams, 1979, p. 252)

The next chapter examines this in more detail and looks at how self-tracking is particularly suited to obscuring neoliberalism – by way of perpetuating an ideology of data-enabled Eudaimonia – in formalised personal science and precision medicine initiatives that have arisen through the quantified self becoming increasingly prevalent in mainstream healthcare.

Chapter 2

The Quantifiable Self: Precision Medicine, the Quantified All' and the Disappearance of Body

> [I]n order to be able to offer each of our patients a course of treatment perfectly adapted to his illness and to himself, we try to obtain a complete, objective idea of his case; we gather together in a file of his own all the information we have about him. We 'observe' him in the same way that we observe the stars or a laboratory experiment. (Sournia, 1961, p. 19)

> Assembling all of this information about an individual from wireless biosensors, genome sequencing, or imaging for it to be readily available, integrated with all the traditional medical data and constantly updated. We now have the technology to digitise a human being in highest definition, in granular detail and in ways most people through not possible, if even conceivable for many decades to come […] algorithms to process it, sets up the ability to track virtually any metric. (Topol, 2012, pp. vi–vii and viii)

An advertisement for the popular activity tracking device Fitbit tells us that it is:

> …on the walk to work, at the weight room or in the last mile. Somewhere between the first tries and the finish lines. Pillow fights and pushing limits. That's where you find fitness. Every moment matters and every bit makes a big impact. Because fitness is the sum of your life. That's the idea that Fitbit was built on – that fitness is not just about gym time. It's all the time. How you spend your day determines when you reach your goals. And seeing your progress helps you see what's possible. (www.fitbit.com)

This is a relatable message. We're busy, our lives are becoming more sedentary, and it's easy to loose track of our diets, health and fitness. We put ourselves at risk by looking the other way when it comes to our health. We need to be vigilant. Being vigilant shows that we care. It means we're doing *something* to be letter, live healthier. This takes work. But, there are pay offs. Technology can

The Politics and Possibilities of Self-Tracking Technology, 35–61

doi:10.1108/978-1-80043-338-020211002

make it easier. If we look inwards with sensors and are disciplined about everything that we do, then we'll be better positioned to reach the goals the health and fitness goals that set for ourselves. We'll be more (pro)active, more resilient against the demands of our increasingly complicated and time poor lives, and more likely to 'unlock' our hidden potential so that we can cultivate a better version of ourselves and we'll have objective numeric proof. Numbers don't lie. If it can be measured, you should track it. The more data you have, the better your results. Or at least that's the promise – if we accept the intensities of measuring every moment, life will get better.[1]

For Wolf this more intense of data-driven approach to measuring and recording things about our bodies, extends beyond fitness and health stands in:

> …contrast to the traditional therapeutic notion of personal development […] When we quantify ourselves, there isn't the imperative to see through our daily existence into a truth buried at a deeper level. Instead, the self of our most trivial thoughts and actions, the self that, without technical help, we might barely notice or recall, is understood as the self we ought to get to know. Behind the allure of the quantified self is a guess that many of our problems come from simply lacking the instruments to understand who we are. Our memories are poor; we are subject to a range of biases; we can focus our attention on only one or two things at a time. We don't have a pedometer in our feet, or a breathalyser in our lungs, or a glucose monitor installed into our veins. We lack both the physical and the mental apparatus to take stock of ourselves. We need help from machines. (Wolf, 2010b)

Wolf qualifies this though, he adds 'watch out for those machines, though. Humans know a special trick of self-observation: when to avert our gaze'. In his *New York Times* article, Wolf (2010b) cites a blog post by Alexandra Carmichael, one of the founders of the self-tracking platform, CureTogether which discusses why Carmichael decided to stop tracking. Carmichael was intense in her tracking.

---

[1]It should be noted that the stand of taking 10,000 steps in a day in order to maintain good health was developed in the mid-1960s from work conducted by Yoshio Hatano a scientist at Kyushu University was investigating the amount of physical activity needed to maintain optimal body weight (Hatano, 1993; Tudor-Locke, Hatano, Pangrazi, & Kang, 2008). Hatano's research was conducted at a time when there was growing concern around the influence of Western culture, particularly diets and more sedentary ways of living and working, was having on the strict disciplined ways of traditional dietary and lifestyle patterns which emphasised health, physical activity and a disciplined approach to everyday life. Hatano's work resulted in the development of an inexpensive personal pedometer by Yamasa Tokei. This device was given the name *Man Po Kei* which translates to English as '10,000 steps'. The release of the device coincided with a period of enthusiasm for physical activity given Tokyo had just hosted the 1964 Olympic Games and sport and athleticism was very much a part of the Japanese public mindset at this time.

She would track up to 40 different variables daily. In the blog post which is in the form of a poem she writes:

> When I first wrote about my self-tracking
> People thought I was narcissistic
> What they didn't see
> Was
> The self-punishment
> The fear
> The hatred behind the tracking
> I had stopped trusting myself
> Letting the numbers drown out
> My intuition
> My instincts
> [...]
> Each day
> My self-worth was tied to the data
> One pound heavier this morning?
> You're fat
> 2g too much fat ingested
> You're out of control
> Skipped a day of running?
> You're lazy
> Didn't help 10 people today?
> You're selfish
> [...]
> Will I ever track again?
> Yes, probably
> For a specific goal or experiment
> Or to observe a pattern
> I'll try to keep an objective, non-judging eye
> But then I'll stop
> When I've seen what I needed to see
> And learned what I needed to learn. (Carmichael, 2010)

However, Carmichael remained active within the QS community though and between 2010 and 2013 she authored over 350 news items on the QS website. Commenting on Wolf's New York Times article, she writes:

> It was somewhat surprising for me to see so many harsh comments [here about the article], until I remembered that new ideas are first ignored, then ridiculed, then attacked, then accepted as obvious. I also wanted to point out that the early adopters here are not just geeks, but also people with chronic health conditions that need constant management. For example, diabetics need to meticulously monitor their food intake and blood glucose measurements - every single day. It's not a choice. Building self-tracking tools that are

> increasingly ambient and passive will not only help people interested in learning about themselves, but will make life immeasurably easier for the increasing number of us who live with chronic conditions. (Wolf, 2010b)

It is this more purposeful form of data capture that makes self-tracking valuable to the healthcare industry. When encouraged data capture through encouraging self-tracking (and sharing) Beyond offering a more complete picture of health and illness for both individuals and populations, self-tracking data affords the construction of large datasets required to develop predictive models and new health-oriented software applications and digital services for both healthcare researchers and consumers. But ultimately, it affords health researchers the ability to conduct experiments outside the constraints of study designs that deductively tests hypotheses and in this sense, research involving self-tracking data can be considered 'post-scientific' because data is captured outside the laboratory by communities who are contributing data into purpose built or already established commercial platforms (such as Fitbit's), or by cooperation between medical researchers and established technology companies such as Apple or Google, start-ups or interdisciplinary research groups who might serve as intermediaries (Swan, 2009, p. 493).

In the context of contemporary healthcare and the industries attached to it, the market for personal data resulting from self-tracking aligns to the restructuring of the division of labour between patients and healthcare professionals. These shifts have a long history but they began to take on a technocratic logic from the 1960s under the auspices of market-centric cost-efficiency measures which includes encouraging individuals to actively participate in their own preventative, diagnostic and therapeutic activities. Here, 'health' is:

> …circumscribed by its standards (quantifiable and measurable, like bodily temperature or blood pressure) and armed with a clear distinction between 'norm' and 'abnormality', should in principle be free from such insatiable anxiety. Again, in principle, it should be clear what is to be done in order to reach the state of health and protect it, under what condition one may declare a person to be 'in good health , or at what point of therapy one is allowed to decide that the state of health has been restored and nothing more needs to be done […] ever new states of body become legitimate reasons for medical or other types of intervention. (Bauman, 2000, p. 79)

And regarding the collection and sharing of personal health data through self-tracking, we could easily argue that argue that society's preconception with 'objective' knowledge about the self draws much of its appeal from popular culture and lifestyle marketing (Wicklund & Eckert, 1992) – both efficient means of perpetuating neoliberal ideologies of health and wellbeing as the Fitbit advertisement quoted earlier demonstrates.

The same can be said about self-tracking technology drawing much of its appeal from the marketing tactics used to promote participatory, and digital, health initiatives. For example, Google Life Sciences collection of anonymous molecular and genetic information sampled from the blood, urine, saliva and tears of approximately 200 people which was termed the Baseline Study (2014). The project aimed to scale up to over 100,000 participants over two to three-year periods and identify biomarkers to better understand disease susceptibility and to develop predictive models that operate at the level of physiological systems – at the blood vessel level to predict the onset of cardiovascular disease for instance. According to the project website the incentive for people to participate and contribute their data is that:

> ...when you join Project Baseline, you're not just participating in research. You have something to offer that no one else can: your unique history, genetics, and way of life. Every data point, every survey answer, and every health test is a way to stake your claim in the map of human health and leave a lasting contribution. (projectbaseline.com)

Project Baseline is just one example of the data-driven research that has been occurring through partnerships between individuals who are, or are prepared to engage in self-tracking, the technology industry and health research institutions. Initiatives such as Project Baseline, are based on an assumption that:

> Intensive testing may uncover indolent disease or incidental findings which, when treated, may cause more harm than good in particular over-diagnosis and over-treatment given that aggressive preventative treatments may not always lead to better patient outcomes, particularly when it comes to prevention. (Diamandis, 2015)

The opportunity that data capture from large cohorts in diagnostic and therapeutic medicine are buoyed by the growing acceptance of self tracking as both a practice that is safe and easy for individuals to practice on their own and as a legitimate means of clinical data acquisition. The empirical models that underpin traditional laboratory and evidence based health research are thus being restructured in ways that conceive of self-tracking systems as part of a broader 'practice bound imagination' (Hyysalo, 2010, p. 78) where innovation in health technology is guided by the objectives of clinical investigators, healthcare administrators, the entities that fund and regulate healthcare research and service delivery. For the technology vendors that cater to this market at an enterprise level, this is a massive opportunity to expand their product and service offerings. When an approach to designing data capturing systems like this is taken, in many cases existing technical infrastructures, components, and data architectures are modified to take into account the needs of clinical investigators. Systems are modified so that they can be configured into integrated platforms which can move data between commercial, clinical and non-clinical contexts (and infrastructures where the data is stored and processed). This capacity to collect data from a variety of sources and in settings that are outside the clinic means that patients have greater flexibility in how they produce and supply data especially when the devices used for capture are not cumbersome, cost effective and integrated into devices used for everyday purposes. For example, Hirsch et al. (2014) have looked at how the *MapMyFitness* application can be used to collect contextual information on physical activity at the level of neighbourhoods over long periods of time (their study ran from 2006–2013). It is interesting to note that in this study the researchers contacted *MyFitnessPal* to obtain datasets, however since October 2013 it has been possible to directly download public data from the *MyFitnessPal* Application Programming Interface (API). Access to this data remains at the discretion of *MyFitnessPal* and could be taken away at anytime. The potential that Hirsch et al. see is in:

> The ability to observe physical activity across large temporal and spatial scales also lends itself to evaluations of policy and environmental interventions to improve physical activity. Researchers could examine patterns of physical activity before and after policy changes at the local, state, national, or even international level. (Hirsch et al., 2014, p. 3)

These researchers argue that there are a number of advantages to using data garnered from smartphone applications through an API. Some reasons being that GPS data is considered to be more accurate than self-reported location data. The same can be said for data gathered through other sensor components compared to self-reported and user inputted data. Given that the data is passive, collection via a freely available smartphone application that functions across a range of device platforms including iOS and Android the data collection burden on both participants and research is reduced. What is more interesting about this study, however, is the recognition that there are also some significant methodological disadvantages associated with the use of passive collection via applications and off-the-shelf devices.

Regarding the generalisability of the data Hirsch et al., note 'users of the application are by definition physical activity conscious, and may not be representative of the general population' (Hirsch et al., 2014, p. 7). In addition to this, it is recognised in this study that smartphone users may be socio-demographically more financially resourced (Hirsch et al., 2014, p. 7) or more technologically savvy. Another interesting effect noted in the study is that the ability to make inferences from the data is compromised due to growth of the dataset in real time which creates a 'moving target' (Hirsch et al., 2014, p. 7). And at the aggregate level there may be noise in the data due fluctuations in the frequency with which participants used the devices though this can be corrected by using algorithm-based prediction of data quality and data reduction strategies to deal with the effects of scale and rapid accumulation (Santini & Romer, 2006).

From n-of-1 to n-of-All

In March of 2015, around three months before Fitbit's would make its first Initial Public Offering (IPO) that raised $358 million US dollars (investor.fitbit.com), then US president, Barack Obama was photographed on multiple occasions wearing a Fitbit surge activity tracking device. This couldn't have been a coincidence. It was a move which spoke to the perceived positive associations between self-tracking, participation in public health initiatives and good leadership, especially among the executive classes where wrist-wear is seen as a symbol of status or success. At around the time of the Fitbit IPO, the office of the Whitehouse was issuing press releases about The President's Precision Medicine Initiative (renamed the 'All of Us' research program which now aims to build a million person cohort), a 215-million-dollar initiative in the 2016 federal budget marked for the development of means to increase patient participation in medical research (see joinallofus.org). The assumption behind the initiative was that if large datasets comprising information that had been sourced from citizens were made available to medical researchers, then the development and launch of new diagnostic tools and treatments could be accelerated and the individuals who contributed their data would benefit from having unprecedented insights into their own health in return. These funds were allocated to the National Institutes for Health (NIH), the Food and Drug Administration (FDA) and the Office of the National Coordinator for Health Technology. This endowment was primarily for the purposes of developing databases, analytical platforms and the regulation and governance frameworks that would be required to ensure the appropriate stewardship of data that had been acquired from the public in accordance with existing standards around security and privacy. The funds were broken down as such: $200m to the NIH with $130m of this going to the development of the cohort and the rest into cancer research; $10m to the FDA and $5m to the ONC.

According to a Precision Medicine Initiative Cohort Program Report from September 2015, the design of a national research cohort that could reach a critical mass of greater than one million participants depended on four key success factors. They were: understanding if it would be best to lever existing cohorts and build on them or start from scratch, or, explore a hybrid model; finding ways to ensure that the rich ethnic, cultural and socioeconomic diversity within the US population was captured within the cohort; having a good sense of what types of data should be captured and what metadata schema and standards should be in place, and what policies needed to be in place, or developed specifically for the initiative for it to deliver on its mission to 'enable a new era of medicine, technology and policies that empower patients, researchers and providers to work together towards the development of individualised treatments' (National Institutes of Health, 2015).

The cohort design project within the initiative was structured by a series of workshops held between April and July 2015. These were conducted to identify the unique scientific opportunities, sort out matters of data organisation and governance (with Vanderbilt University), understand how to ensure participant engagement and equity and understand the technical issues (with Intel). The workshops were followed by requests for information and surveys about public perceptions of the proposed data collection and governance models. One of the outcomes of this work would be the establishment of privacy and trust principles. In the survey 79% of respondents supported the concept of the cohort and 54% stated that they were likely to participate. Of those who said they would contribute data to the cohort 82% were interested in receiving study results and personalised analytics based on their data and 62% said that they wanted to help advance health research. 71% of respondents said that those who contribute data should be acknowledged as research partners.

In one of the public consultations calling for feedback on the NIH's proposal for a the cohort raised five considerations that were reported on: (1) the willingness of participants to carry their smartphone or wear wireless sensor devices sufficiently throughout the day so that researchers would have complete enough data for them to be able to adequately use in their studies; (2) the willingness of participants without smartphones to upgrade to a smartphone at no expense to them if one was offered; (3) how often people would be willing to let researchers collect data through the devices without it being an inconvenience; (4) the kind of information that participants might like to receive back from researchers and how often; (5) alternative forms of collection that might not rely on smartphones and wearable devices.

The workshop on scientific opportunities identified a number of areas where the one million person cohort might enable significant developments in certain areas of health research. They were: the discovery of new biomarkers for the prediction of future disease risk in individuals; the discovery of quantitative risk determinants of individual variation in response to certain therapeutic interventions; quantify the population level risk to certain diseases by integrating data on environmental exposure, genetic factors and their interaction; the integration of mobile health (mHealth) technology into better functioning systems; discover new classifications and relationships between diseases and enabling target clinical trials with subjects with rich clinical data and making this data more broadly and openly available to investigators and other research partners. The cohort was conceived with participation in overall governance, dissemination of findings and evaluation of its effectiveness. Consent was framed as operating at the levels of basic participation, secondary use, when new components and measurements are proposed and for the option to join new supplementary and complementary studies. When data is disseminated or returned this was deemed to be according to individually set preferences on the sharing and return of information and that at minimum aggregate results should be available to all participants. An example of what the data sources within the cohort, as reported to the advisory committee director is shown in Table 2.1.

When traces of everyday activity are recorded, or certain physiological variables are tracked on an everyday basis, they are often also submitted to online platforms where analytic techniques developed by third parties are used to produce a synthetic view of the individual which is then fed back to the user in the form of insights. At the same time, this information is afforded a type of portability that extends beyond the purpose for its primary collection. When shared with other intermediaries in de-identified, aggregate or enriched form, this all the types of data listed in Table 2.1 it is not hard to imagine the extent to this data are 'gathered under conditions of profound power imbalance' (Nissenbaum & Brunton, 2013, p. 171). Placing these conditions and the power (a)symmetries

Table 2.1. Possible Data Sources[a] Imagined for the Precision Medicine Cohort.

Data Source	Type of Data Provided
Self-reported[b]	Diet, substance use, disease symptoms, mood, family history
Structured clinical data (EHR)[b]	International classification of disease (ICD) codes, medication history, laboratory results, vitals and encounter records.
Unstructured clinical data (EHR)	Narrative documents, images, EKG and EEG data
Biospecimens[b]	Blood and plasma samples, genetic data (from saliva), environmental exposure (from hair and nails), microbiome (from stool)
mHealth technology and digital sensors[b]	Passively-collected data (location, movement, social connection), wearable sensor data (activity, calorie expenditure, sleep, physical activity)
Health insurance claims	Billing codes (public and private) pharmacy dispensing
Geospatial and environmental	Weather, air quality, pollutant levels, population density
Other (cross referenced c.f. self -reported)	Social network data, over the counter medication purchases

[a]An EHR is an Electronic Health Record.
[b]denotes that this source forms part of the minimal core dataset.

associated with them in a critical context requires examining a number of processes that bring data into existence and operationalise it for its various uses.

As observers of these data acquisition initiatives and their relation to the subject as consumer of healthcare – a body that requires or may require in the future, diagnosis, therapeutic intervention or quantifiable prognoses and also a body can be mined for information (i.e. a *quantifiable self*) – we can identify two overlapping modalities in how data from self-tracking is refashioned for third parties and used to derive value (see: Couldry & Mejias, 2019). The first, 'technical transience' refers to a celebration of technology among participants and advocates of self-tracking but also an ambivalence towards technology as well. The technology used to capture and analyse self-tracking data is open, replaceable and amenable to constant refinement. Second, the complete cohort system comprises of objects that collect data, the platforms, information schemas and analytical methods that enable that data to be transformed into information. This incorporates symbolic apparatuses that are designed to legitimate the use of data-garnering objects, encourage their continual use and compliance. This refashioning also has an impact on the ways that bodies are represented within clinical ontologies. For example, the power symmetries embedded in the data infrastructures that turn seemingly raw data into meaningful insight in a way that, as Eugene Thacker suggests, biological components and processes are technically re-contextualised in ways that need not necessarily be biological resulting in novel configurations of the biological and technological that describe an ambivalence that is not reducible to either a 'rhetoric of enabling technology' in which the 'technology appears to disappear altogether' (2004, pp. 5–6). And at the same time the individual body dissolves into data.

Traditionally, the production and stewardship of clinically relevant data was restricted to healthcare professionals with the exception of those monitoring blood sugar or pressure for the management of chronic conditions such as diabetes or hypertension. But the way that the Precision Medicine Initiative Cohort project frames 'patient data work' (Piras, 2019) – the work of gathering, storing, sharing and interpreting both data and the protocols around its capture and re-use – participants are 'prosumers' (producers and consumers) of their own health data. As Enrico Piras argues,

> …the conceptual landscape [of health data gathered through self-tracking] is confused because there are several competing labels that aim at describing patient data and because these are used interchangeably despite their original meanings. (Piras, 2019)

The disambiguation offered by Piras is helpful in understanding the basis of power asymmetries with respect to health, the producer-consumer relationship, the status of data and the division of labour involved in its production and use of data that would be captured by the cohort project. This is summarised in Table 2.2.[2]

[2]Overall, a medical history is a chronological narrative of past events and circumstances that may be relevant to a patient's state of health that is gathered ideally first hand from the patient by direct questioning from a healthcare professional. Clinicians are trained to ascertain the components of a medical history by seeking out the following: Chief complaint (CC) a brief statement of the complaint or incident that prompted the patient seeking care; History of Present Illness (HPI) a detailed chronological account of the development of the current health issue from onset to the present time in the patient's own words; Past Medical History (PMH) prior illnesses, their treatments and sequelae; Social History (SH) relationships, occupations past and present, travel, hobbies, stresses, diet, use of alcohol, tobacco and drugs; Family History (FH) present health or cause of death of parents, siblings and hereditary disorders in family; Review of Symptoms (ROS) an exhaustive survey of symptoms or diseases organised by body system not covered in previous parts of the history-taking. The protocol for the order of body systems examination is: Cardio-vascular, respiratory, alimentary, central and peripheral nervous systems (including psychiatric health), the genital and urinary systems, locomotor and musculoskeletal system, endocrine system and the skin. Beyond this, details about doctors that the patient has seen, hospital and clinic visits and alternative healthcare practitioners that the patient has seen are queried. Heteroanamnesis involves gathering information from the patient's family members or others close to them.

Table 2.2. Mechanisms of Health Data Capture.

Framing of Data Capture Mechanism	Characteristics and Approach to Stewardship and Interpretation
Patient Generated Health Data (PGHD)	Anamnesis and heteroanamnesis, biometrics, treatment history, lifestyle information that is created, recorded or gathered from patients or their delegates to address a particular health concern.
Observations of Daily Living (ODLs)	Patterns and realities of everyday life traditionally not thought to be essential information such as data about mood, sleep, physical activity. ODLs are usually collected by patients using tools of their own choice and they exercise their own discretion over what they share with healthcare providers. A barrier to ODL data collection is the burden it places on patients and it is here that passive digital technology applications might apply.
Quantified Self (QS)	Self-experimental data gathered with a specific problem in mind and collected with a desire to manipulate factors to achieve desired ends. These ends can sometimes align or deviate from clinical objectives, for instance, tracking to not over medicate.
Personal Health Information Management (PHIM)	Less emphasis on specific data types but rather on social and cultural factors that support the access, organisation and integration of health data. It is within this frame that many consumer-facing devices and applications are designed.

Source: Adopted and developed from Piras (2019).

These mechanisms of data capture are not new and have been part of the agenda for technology applications in healthcare since the advent of computational technology. For example, in his essay, *As We May Think* (1945), Vannevar Bush mapped out an agenda for post-war science and R&D where the chief objective was the task of unlocking and making accessible, and interoperable, the repositories of warfare related scientific knowledge in times of peace. Bush wrote:

> Of what lasting benefit has been man's use of science and of the new instruments which his research brought into existence? First, they have increased his control of his material environment. They have improved his food, his clothing, his shelter; they have increased his security and released him partly from the bondage of bare existence. They have given him increased knowledge of his own biological processes so that he has had a progressive freedom

> from disease and an increased span of life. They are illuminating the interactions of his physiological and psychological functions, giving the promise of an improved mental health. (Bush, 1945)

While for biomedical sciences the war and peacetime research agendas remain relatively consistent – with the exception perhaps of research into bioweapons – the scale and interoperability of information held great promise for the development of new biomedical technologies which took their inspiration from outside the traditional domains and lineages of the existing medical technologies of the age. Bush wrote:

> The great advances in medical science have not originated in the past to a sufficient extent in the research of medical men themselves. They have come more often from the chemists, the bacteriologists, the physiologists, in the universities and in the pharmaceutical industry. They have arisen as by-products from such unexpected places as the dye industry …The primary contribution of medical men has usually been in clinical research … or research in surgery or the like, utilising products and instruments that appeared outside the profession and aimed originally at other purposes. There is a gulf between medicine and other professions, widened at times by the instance of the medical man upon his prerogatives, and crossed only too rarely by medical scientists of true breadth and vision. (Bush, 1945)

While a medical version of the memex remained largely unrealised its essence is retained in a range of health and medical technologies ranging from: the use of computer terminals in clinics and hospitals to automate parts of the patient intake and medical history taking processes under the main imperative of cost/time reductions; the digitisation of medical information and literature in databases and searchable websites such as WebMD, PubMed; the use of standardised information taxonomies and metadata schema such as the National Library of Medicine's hierarchically-organised vocabulary for indexing and cataloging medical information, MeSH, the personalised health network and online community, Patients Like Me where users share real world experiences of their conditions and treatments, and AI systems which use crowd sourced data and machine learning for end-user self-assessment via smartphone, Ada.[3]

If we take a process-oriented view of how decisions are made about an individual's health by doctors and allied healthcare workers or by care-givers and families in the course of treatment or in addressing the aspects of an individual's

[3]Since the 1960's the use of computers has been prototyped to develop closed loop medical interviewing systems designed to automate, the task of taking medical histories and make it more consistent (see Peckham, Slack, Carr, Van Cura, & Schultz, 1967). Interestingly, among Slack and van Cura's findings was the idea that 'patient-machine rapport was quickly established [and patients] earnestly participated in the interview[s]' (1967).

lifestyle which might exacerbate or prevent disease, personal data becomes a way of addressing uncertainty associated with other variables. More readily, what we observe in data-driven medicine is an attempt of information capture where: (1) mathematical and algorithmic modelling are used to mediate, and predict physiological phenomena; (2) software affords the ability for investigators to conduct experimental work outside the constraints of deductive work testing hypotheses; (3) medical experimentation becomes post-scientific in a way that sees innovation occur in spaces outside the laboratory and increasingly in partnership with established technology companies, start-ups, coalitions small interdisciplinary academic teams, entrepreneurs and citizen-led groups such as the Quantified Self (Swan, 2009, p. 493).

Like many of these developments, data capture through self-tracking is also a departure from what usually happens in traditional forms of clinical research where data is collected with narrowly defined research questions and hypotheses in mind so that the messiness of contextual and other factors would not contaminate the data. For example in Randomised Controlled Trials (RCTs) complex cases where a patient might present with multiple conditions are treated as outliers even though they are likely to typify the patient profiles that working clinicians would encounter in their practice. According to a press release issued on 28 July 2016 by Fitabase, a platform which claims that it is 'enabling researchers to use the latest tools, devices and apps to further collective knowledge [and make] it easy as possible to deploy new tools of wellness to measure, track and engage [study] participants' (Fitabase, 2016), there were 200 clinical studies that were using data garnered from Fitbit devices. According to a statement attributed to Fitabase CEO Aaron Coleman (Fitabase, 2016) there are over a billion minutes of Fitbit data on the platform which had been collected on behalf of the company's customers in the research sector.

The platform also provides its customers with a 'knowledge base' where different Fitbit products can be compared for assessment of their suitability to a particular clinical trial or experiment. This knowledge base also provides a range of informational resources designed to help researchers with working with internal review boards to get projects using Fitbit technology approved. This includes scripted answers to common review board questions on data security and privacy along with guidance on how to work with research participants and templates for consent forms for use in their studies. Through its fee-based Application Programming Interface (API) service, Fitabase allows researchers to connect data from the platform to Electronic Health Records (EHR), export and share data with other systems they might be using, and communicate with research participants by text message or email, presumably by a numerical identifier without researchers seeing participant contact details. The platform offers a research library linking to a corpus of published peer-reviewed articles where a Fitbit device has been used in an experimental study. At the time of writing there are 761 publications in this library. As the use of third-party data becomes more prevalent in health research, models for documenting what happens to data from the point of acquisition to analysis becomes an increasingly important consideration in research design and research ethics.

Data Provenance

At the European Quantified Self Conference, a meeting for those participating in n-of-1 personal data, Andreas Schreiber has proposed a provenance model for health tracking data. Schreiber's model defines provenance as 'information about entities, activities, and people involved in producing a piece of data or thing, which can be used to form assessments about its quality, reliability or trustworthiness' (2016).

According to Ford and Graham, the problem is that 'as digital data becomes abstracted into short data statements that can be shared according to the logics of 'linked data', the concentration of power' (2016, p 958) of data controlling institutions is increased. For Ford and Graham, increased control is granted to institutions by a 'loss of provenance' (2016, p. 958) and for these authors the object of provenance is source information or metadata about the thing that a piece of data represents. The model is summarised and developed in Table 2.3.

This three layered-provenance approach contributes to understanding what the tools that have been used to generate these data are and how they are being used. It considers the exact mechanisms by which these data are generated. This is a question of experimental method and system design. When data is mobilised in the ways outlined earlier in this chapter (Table 2.1) it becomes difficult to trace and report back to participants. The various data sources become difficult to detangle. Disentanglement and being able to re-trace and reverse engineer data-motile processes and transactions where data sharing or linkage has taken place are important given that often data capture systems are formed through relations where certain parties (e.g. data custodians, research scientists, clinicians) are not particularly concerned with future developments or implications of the

Table 2.3. A Three Layer Provenance Model.

Object of Provenance	Questions
Entity	What data about the subject were created during the capture?
	What data about the subject were automatically generated?
	What data about the subject were captured by manual input?
Activity	Which activities support visualisation, analysis and interpretation of the subject's data?
	In which activities can the subject input or modify their own data?
	What processes (online and offline) are communicating data?
Agent	What parties were involved in generating the data?
	What parties have or are requesting access to the data?
	Can other parties see the subject's data?

Source: Adapted from Schreiber (2016).

technology given that the ethical responsibility, burden of care is beyond the purview of contractual, service, or support agreements between the 'owner' of the data and the those who design and implement a working system – the intellectual property here is not in the data but in the technology built over the top of it; the data may have been used as a one-off purchase or provided under licence. The interests of users and third-parties beyond the data owner, custodian, or subject (of the data) have 'skin in the game'. And although some rights and control may be retained (by data owners, custodians or subjects) over data assets and they may even get a say in the terms under which sharing sensitive information occurs, it is a limited, conditional and probably unsustainable form control. More often than not third-party interests are hidden, downplayed or sold as harmless or even beneficial and the terms of data acquisition and sharing arrangements go unquestioned. Part of this has to do with the sophisticated rhetoric about privacy that's being used by the major technology companies when promoting their various data-driven healthcare initiatives. For example, in 2014 after the health application and developer platform allowed applications that monitor indicators of health to be housed in within the device for sharing with hospital software systems or directly with individual healthcare professionals. In an interview with Bloomberg's Charlie Rose, Apple CEO, Tim Cook trying to sound reassuring, stated:

> This is another way to begin to build a comprehensive view of your life which should empower you to take care of yourself over time and when you need help it empowers you to take certain data to your doctor to get help from them all the while guarding your privacy so that nobody's getting the data if you don't want them to have the data, nobody's sharing the data if you don't want them to share the data [...] and no we're not keeping it. (Bloomberg, 2014)

Despite such assurances, when personally collected data is brought into encounters with healthcare institutions via the technology industry, the involved parties enter into a relationship with human bodies where there is a tension between the need to deny the body its separation from data in order to be able to produce the level of data-granularity required to offer 'objective precision care' and at the same time produce sufficient equivalency between data and the body in order to sustain it as something that has value in the form of a commodity. Eugene Thacker argues that the production of medical facticity through processes of digital mediation (of which data capture is one) is a 'discourse of the body, and a discourse that articulates specific kinds of bodies framed by physiological, genomic or biochemical markers' (Thacker, 2004). The relationship between the body appearing and disappearing through datafication, and the presence or absence of markers that are additionally becoming functionally non-distinct of digitally encoded forms. For Thacker, mediating the body is a process in which biological components and processes are technically re-contextualised in ways that need not necessarily be biological (e.g. mounting slides in microscopy, the conversion of sound to images in sonography or data aggregation/reduction in large cohort studies).

The integration of personal health data across a range of sources such as whole genome sequencing, clinical laboratory tests, user inputted data and self-tracking data from standalone devices or mobile operating system health data APIs in order to produce baseline data around wellness as opposed to tracking chronic disease metrics also raises issues. Recently, Google introduced Google Genomics, an API that allows for genomic data which is usually too large to download and perform analysis on without dedicated computing infrastructure to be stored in the cloud and moved into Google's server farms for analysis. As the data from more and more public genome projects are added to this database, and at present it is reported that there are 3,500 genomes available on the cloud there is likely to be a boom in genome data oriented startups who will build products and services drawing from this data asset and integrating this data with other sources of data to develop services and solutions across a range of industries.

To better understand this, we need to:

> ...jump straight ahead to connect the vast arrays of life and history, to mobilise gigantic forces, to detect dramatic patterns emerging out of confusing intersections, to see everywhere in the case at hand and yet more examples of well-known types, to reveal behind the scenes some dark powers pulling the strings. (Latour, 2005, p. 22)

Homo Economicus

The power dynamics described in this chapter invite the question of how we might continue to read Foucault – as many scholars in this area do – as being relevant to the analysis of self-tracking. Especially his later works on subjectivity (technologies of the self), bio-politics as a 'powerful form of governmental reason' (Dilts, 2010). Given that Foucault took particular interest in the reconfiguration of quotidian practices in the neoliberal analysis of human capital, 'a revolutionary re-description of all activity as forms of personal investment, all consumption as production, and rendering all actors as entrepreneurs of the self' (Dilts, 2010).

One of Foucault's accounts of political subjectivity is given as 'the surface of contact between the individual and the power exercised on him and so the principle of the regulation of power over the individual, will be only this kind of grid of *Homo Economicus*. The figure of *Homo Economicus* is the interface of government [or control] and the individual' (Foucault, Davidson, & Burchell, 2008, p. 226). The grid structure mentioned here comes into sharper focus when technology becomes part of this interface. If neoliberalism is to be thought of as the embodiment of rational choice (Dilts, 2010) then the precision of sensors and the seemingly objective capacity of data to be analysed only makes the choice to embody logics of self-governance more rational.

This is an 'enabling subjectivity' (Dilts, 2010) that allows for the tools that are used to measure environmental, biological and behavioural phenomena to be applied to nearly all aspects of social life – education, work, economic productivity, social inclusion and cohesion – and attaches to it an indexical notion of potential to clinically relevant. A rational actor:

> ... responds systematically to modifications in the variables of the environment, appears precisely as someone manageable, someone who responds systematically to systematic modifications artificially introduced into the environment. *Homo Economicus* is someone who is eminently governable. From being the intangible partner of *laissez-fair Homo Economicus* becomes now becomes the correlate of a governmentality which will act on the environment and systematically modify all its variables. (Foucault et al., 2008, pp. 270–271).

The self-tracker as a *Homo Economicus* like figure offers a way to think about the power symmetries embedded within the relations forming between bodies as biological, political and informational entities. Foucault's account of *Homo Economicus* contributes to the illumination of techniques, objects and modes of power flowing through the use and design of self-tracking technology as both material and psychological 'technologies of the self' beyond simple accounts framed by the jurisdictional power of medical elites (c.f. Jewson) and brings the lens closer to areas where the boundaries between living a good life (eudaemonia) and an industrious (neoliberal) one are not so clearly drawn.[4]

For Drew Leder there is a difference between awareness of the body fading into the background, when things are 'well' the body 'dis-appears' and when the body is unwell and experiencing pain or illness (1990a, 1990b, p. 84). The processes that healthcare professionals use to make the sick or unwell body knowable 'seek to soften bodily dys-appearance and enable bodily dis-appearance' (Zeiler, 2010, p. 333).

---

[4]The eudemonic conception of wellbeing holds that well-being is not an outcome or end-state that can be reached but rather is an ongoing process of realising potentials. The eudemonic conception of wellbeing is based on six characteristics: self-acceptance, personal growth, relatedness, autonomy, relationships, environmental mastery and purpose in life (Ryff, 1989, 2017; Ryff & Singer, 2006). Bauer, McAdams and Pals (2008), note that people whose life-narratives are rich in intrinsically motivated goals for personal growth or those that display concerns for integrative growth based on the integration of old and new perspectives derived from a deep self-understanding will report higher levels wellbeing especially around indicators of personal growth and purposefulness.

The blurring of this line starts with the assumption that individuals have the power to participate in healthcare and that in doing so they benefit in non-monetary ways where the foregone value is only that of the time and effort (Becker, 2007) taken to track, measure and share their data. The application of Foucault's thinking on technologies of the self and *Homo Economicus* do lend themselves to the study of mainstream, consumer facing self-tracking particularly given neoliberalism's 'aim of producing free and autonomous individuals concerned with cultivating themselves in accord with' (Hamann, 2009, p. 48) the vision of a quantified self. However, this approach does have its limitations, particularly when we are seeking to understand power, and political subjectivity, beyond the institutional governance of individual bodies and extend it to the interface between bodies and institutionally governed technical and informational logics that contribute to the embodiment of neoliberalism. This embodiment may hinge on the design of the objects and systems that make self-tracking possible to the mechanisms that incentivise it as a sustained practice of cultivating and caring for the self. Though a Foucauldian analysis of these relations between self-tracking practices as sustainer of neoliberalism is productive to an extent, it is Foucault's explicit rejection of dialectics that can be limiting – unless one seeks out and builds on the kernels of dialectical thinking in his work (see Grant, 2010).

Although I will not go so far as to suggest that Foucault can be read as a dialectician compared to Herbert Marcuse for whom for example in his essay *Some Social Implications of Modern Technology* (1941/2004) saw potential for technology to be both oppressive and liberating at the same time, seeking out a blueprint for dialectical thinking in his account of biopolitics is necessary if his work is to be read together with proponents of dialectics and dialectical materiality as I will be doing in the following chapters. This way of reading Foucault, as John Grant argues,

> ...gives recourse to Foucault for those who see dialectics as a requirement of radical politics, while also providing a platform for future research that reconnects the study of power relations with dialectical themes such as experience, liberation and ideology. (Grant, 2010, p. 220)

Foucault's focus on genealogical histories brings to light real instances of dialectical tensions in the visions for certain technologies in terms of their potential to simultaneously empower and disempower.

For example, in his essay *Nietzsche, Genealogy and History* (1978) Foucault explains how the body is

> ...molded by a great many distinct regimes; it is broken down by the rhythms of work, rest, and [leisure]; it is poisoned by food or

> values, through eating habits or moral laws; it constructs resistances. (Foucault, 1982, p. 87, as cited in Grant, 2010, p. 227)

If a blueprint for dialectics can be found in Foucauldian thinking then it can be applied to account for a number of interpretations of power without denying an emancipatory politics, which in regard to the argument being advanced here, is one of a radical politics not directed at macro bio-political subjectivities of neoliberalism but one of micro-resistance(s) situated squarely in the relationship between experimental and system design and use and the perceived and realised value of patient participation in healthcare.

Following Foucault, Nicholas Jewson sought to critique the political and economic forces underpinning the production of medical knowledge. Jewson offers two explanations for the way that bodies are abstracted when the markers of sickness (and health) are aggregated, categorised, and documented in the modern practice of healthcare. This happens when a subject, the patient, is reduced to a set of signs and symptoms and then secondly when these symptoms and their perceived causes are pinpointed to various organs and systems. Within this particular information geometry, lay-knowledges are posited as an "independent [...] enterprise able to address the interpretative challenges that the edifice of modern medicine poses to the sociologies of knowledge and power' (Nicolson, 2009). The political ontologies (cosmologies) of institutionalised healthcare, Jewson notes that they:

> ...may be seen not merely as cultural artefacts, lying outside the realms of social discourse, but as modes of social interaction within the structures of relationships which surround the production of medical knowledge. It is contended that [...] cosmologies generate, reflect and project conceptions of order and identify in the network of relationships that constitute the process of innovation in medical knowledge. They function as a medium within and through which perceptions of the self and others are expressed, legitimised, and institutionalised. (Jewson, 1976)

Jewson's argument is as follows. There is a transition from bedside to hospital to laboratory-based medicine and through this transition the sick-person 'disappears' as both a target for therapy and as a subjectivity. The sick person, as patron and subject in the healthcare industry becomes an object of medical knowledge but not in the way that it may have traditionally been involved in the labour processes associated with the production of therapeutic outcomes. Medical investigators are the group with autonomy over the production of medical knowledge and engage in the material labour of its production and distribution. Healthcare, as a system of patronage, binds patrons, medical scientists and clinicians and dominant epistemological paradigms of the day dictate the conditions under which these relations are structured.

The raw materials of production are essentially information about the body and its environment, – its material composition and properties, its processes (represented) its observable characteristics, and that which can be expressed in language or gesture (by the subject). The process of producing medical knowledge – the predominant activity of many medical professionals – is carried out by way of applying theoretical and empirical techniques that fall within certain conceptual frameworks and widely adopted experimental and therapeutic guidelines. The product of this 'system' is a set of theoretical presentations and courses of action where the subject is modulated between the position of being an individual case and part of a population-based cohort simultaneously.

In a set of commentaries on Jewson's original text in the International Journal of Epidemiology readers are reminded of that as healthcare becomes more technologically contingent and governed by neoliberalism, healthcare is accompanied by, though not exclusively, the potential for individuals to have greater autonomy, influence and informed engagement with medical professionals and healthcare institutions. But at the same time these shifts introduce 'inequalities, excesses and vagaries of a system structured by unregulated market forces' (Nicolson, 2009). Given Jewson's account was one that referred to theoretical principles of the medical guild and attempted to metaphysically describe the production of medical knowledge and discourse (including: concepts, models, statistics, diagnoses, prognoses) we can think of cosmology in a more contemporary sense as a medium or platform through which the production of clinically valuable data is commodified (see: Tables 2.4 and 2.5).

Table 2.4. Modes of Health Knowledge Production.

Location of Data Collection	**Patron**	**Role of expert(s)**	**Subject**	**Objective of Data Collection**	**Illness Framed as**
Bedside	Patient	Healer	Person	Prognosis, treatment	Whole person
Hospital	State	Clinician	Case	Diagnosis, classification	Organs, physiological systems
Laboratory	Researchers	Scientist	Cells	Analysis, explanation	Biochemical processes
Digital sensors	Industry	Analyst	Data	Aggregation, value creation	Correlation between variables

Source: Adapted and developed from Jewson (1976, p. 624).

Table 2.5. Evolution of Medical Ontologies.

	Bedside	Hospital	Laboratory	Data
Logic of classifying disease	Totality of symptoms	Internal events in organ systems, causal	Cellular function	Correlation between quantifiable variables
Focus of pathology	Systemic	Localised	Biochemical and physiological processes	Indicators of disease or abnormality (or potential for these things)
Research methods	Situated observation, speculative inference	Statistically oriented clinical observation, cases documented and archived in files	Laboratory experiments, scientific method and peer reviewed publication	Real-time data collection, self-reporting using devices, passive ongoing extraction of data
Diagnostic technique	Qualitative judgement	Physical examination	Microscopic, in vitro, and by bioassay	Mathematical, computational and mediated
Therapy	Heroic and extensive	Sceptical (with the exception of surgery)	Nihilistic	Optimistic, preventative
Mind/body relation	Integrated as same pathological system	Differentiated, psychiatry is a specialised area of practice	Differentiated, psychology is a specialised area of practice	Neuroplastic and neuro-diverse

Source: Adapted and developed from Jewson (1976, p. 624).

In light of Tables 2.4 and 2.5 above,

> …[the] responsibility, and increasing culpability for health status lies with the 'person' who is impelled to become aware of, and act upon, the wealth of health information and advice to be found not only in the […] clinic, but throughout various forms of popular media, on supermarket shelves [and] on the internet. (Nettleton, 2009)

Increased layperson participation in medicine extends the laboratory medicine paradigm transforming it most markedly in terms of how it accommodates data produced and shared. The integration of patient/consumer data alters the rate and extent of processes of change within and between modes of production in healthcare.

The production of data about the body though self-tracking reorients the emphasis clinical inquiry back onto the individual, transforming laboratory medicine in a way that transforms individuals 'thereafter into independent [producers] and consumers [of medical knowledge] in [the] market-based systems' (Prior, 2009) of a particular moment in history where we are seeing increased technocracy in medical research.

> [Jewson] illustrates how the 'sick-man' is not just an idea; a concept; but a figure that arises out of the socio-economic relations between physicians and their (wealthy, independent) patients who can demand that they be treated as unique individuals. In the same way, the patient as a 'case' arises not merely out of new ways of thinking about disease, but structured through a new set of relations – focussed on hospitals as a place of study and treatment in which an emergent cadre of salaried, professional, independent physicians come to view human organs rather than human beings as their raw materials. Similarly, it is the material base of the laboratory that is connected to the production of science and further [a] new cadre of salaried scientific researchers who can drive medicine and medical practice forward in accordance with their interest (in cells and germs) rather than those of the clinicians, – or indeed of the patients. Henceforth, the patient becomes not merely the subject of a professional clinical 'gaze', but also of the apparatus and enterprise of 'medical science'. (Prior, 2009)

The focus here is that health risks re-orient the clinical gaze towards, prevention, screening and self-management all of which have the economic benefit of reducing the resource burden on the healthcare system generally but also on the burden of acquiring clinically relevant data from populations. As Prior argues:

> ...the modern sick-man and woman are no longer viewed as 'patients', but [rather] as independent consumers with responsibility each for their own health, and [are] expected to express that responsibility through choice and action in the healthcare market, it is not ... because of any new cosmology. Rather, it is an expression of everyday material circumstances in which we all live, and [in which] medicine is practiced. (Prior, 2009)

One direction for this line of thinking is to consider how self-tracking contributes to the figuration of an industrious body that is made demonstrable, to the self and to others through representations that are 'imbricated' by this exact lens of neoliberalism. On this point, Saskia Sassen points out

> [...] the term imbrication [captures] the simultaneous interdependence and specificity of each the digital and the non-digital. They work on each other, but they do not produce hybridity. Each maintains a distinct irreducible character. (Sassen, 2008, p. 345)

The reification of physiological processes, geographic coordinates, timestamps and behavioural routines in this linked way that retains the non-reducible construction of the body:

> ...is a historically specific mode of organisation of material forces, invested by capital into being, as well as elaborated through various discourses of biology and physics, thermodynamics and complexity, metastability and nonlinear rationality, reconfiguring bodies, work, and reproduction. The bio-mediated body is a definition of a body and what it can do – its affectivity – that points to the political, economic and theoretical investment in self-organisation inherent to matter or matter's capacity to be informational, to give bodily form. (Clough, 2008, p. 2)

This hybridity of bodies and data brings with it the introduction of new mediated experiences of selfhood and subjectivity and an influx of new and conflicting streams of information and decreasing relevance of institutional and social collectives (the family, peers, education system, employers, the state). In everyday decision making has resulted in the creation of the feeling that there is an unprecedented ontological complexity associated with life and in order to 'succeed', an individual must learn to cope at all costs. It also validates the idea that living in this complex world requires a self-directed effort of compiling information and resources, planning and execution. At the same time this process involves avoidance and cognitive dissension, denial, refusal of certain things that are actually or potentially known and that do not fit harmoniously with the objectives of one's life-project; The intertwining of feelings of alienation, powerlessness and blindness for which self-tracking technologies offer a means to take control and know certain things about oneself in a way that affords comparison to others.

In this 'imbricated' form, the body is framed by contradictory forms of knowing. Personal, intimate and somatic knowledges are packaged up and fed back to individuals as a neat, combined narrative amounting to a 'full picture' of one's life that is defined by consumer culture, lifestyle marking and the healthcare industry (see Giddens, 1991, pp. 187–201). The collection of personal data tempts participants with the sense of security and empowerment that comes from being able to know and see all. But this data-driven narrative is blind to the deeper, hidden or obscured pathologies that are overshadowed by neoliberal framings of health – and self-determination – being linked to productivity and wellbeing. For example, data that suggests a state of good health based on heart rate variability says little if anything really meaningful about someone's vascular health or stress management; similarly data about daily hydration levels says very little about the health of the digestive system or the kidneys but both these things are readily measured and used as generalised indicators of 'health' by a range of devices and applications available on the market at any given time.

This reflected in an exchange between a Fitbit user and a customer service representative on the company's online community forum, the user raises a query about how the device automatically sets the daily calorie goal once user data on height, age and weight is entered. In this instance, the user was suffering from a chronic illness which makes it difficult to absorb nutrients so their motivation to self-monitor was not to lose weight but to gain and maintain it. The user reported that their daily calorie goal had not been adjusted by the device despite the user profile being set to maintain rather than lose weight. Assuming that the calorie burn goal is calculated by a device-specific, proprietary formula, the user states 'I can't imagine that it is easy for those who question, and challenge, excessive calorie burn goals when they are presented with them'. Writing on this type of tendency with the logic of self-tracking devices and software architectures, Benjamin Bratton notes in the chapter dedicated to self-tracking titled 'the quantified self and its mirror' in *The Stack*:

> The baseline design brief for the user layer of the stack may go something like this: In the image of planetary scale information infrastructure, comprising trillions of addressable necessities, the resolved scale of the platform need not be for one user at a time, drifting into and out of narcissistic virtual reality, but for pluralities of partial users, quasi users, human and non-human organic and inorganic intermingling in intersubjective alliances predicted not just on autobiographical interoperability but on direct physical and cognitive promiscuity [that is projected onto it]. (Bratton, 2016, p. 264)

As the conversation continued between the user and a customer service representative, the user points out:

> The specific issue I raised [is] that there's a glitch [with the software] the calorie burn count doesn't go down when weight and BMR go down. Why? This isn't just about me. Health and fitness are not intrinsically tied to weight loss. The word maintain does not mean remain stable or drop – it means to remain stable and stay the same. Not everyone in your community has the same health goal or needs (there are plenty of people who want to be active for whom weight loss is dangerous or simply unwanted) and I would love to see Fitbit increasing its cultural competency around this issue or not pretending that they, as a company, are competently providing goals for anybody except those who hold weight loss/attaining an idealised body type as their primary goal. The majority of average folks will assume they can trust Fitbit's

> algorithms to do what they say they do, within a reasonable margin of error, and Fitbit doesn't provide its math for people to check or adjust. (Fitbit, 2016)

This brief exemplar shows us how those participating in self-tracking being constructed as a 'generalised type of user-agent within a diverse throng of alternatives' (Bratton, 2016, p. 261). The user-subject is at once synthesised and made alien to itself by the experience of self-tracking being reflected back through the logic of visual-quantitative technologies; and a subjectivity being projected on the non-human representation of human experience (i.e. data to be acted on). For Bratton, the 'death of the user' here constitutes the 'displacement of its soft humanism from the conceptual centre of design strategy by the proliferation and pre-dominance of both non-human and non-individuated actors' (2016, p. 391). This version of the self is one where the nature of the body is transformed across the axes of individuals/populations, time/space, and interiority/exteriority (Turner, 1984, p. 82). This is not just about users becoming crystallised as a layer within a technology stack. It should be more about understanding, and designing for, the ways, both scripted and non-scripted, expected and unexpected, that user-subjects negotiate their position within this layer. This type of interaction does not have:

> ...a stable division of labour corresponding to an unproblematic partition of the ... human body. New specialties are created around overlapping principles: organ systems (e.g. cardiology, ophthalmology), occupational criteria (e.g. sports medicine, industrial medicine), and so on. The heterogeneity of principles does not mean, however, that biomedical practitioners can arbitrarily establish divisions within medicine and biology in order to create distinct and independent domains of professional expertise. The independence of biomedical speciality is inevitably relative: haematologists, for example, cannot ignore pathology reports when it comes to treating patients. In medicine this mutual dependency is stronger than in most professions, such as engineering, since the object of medicine is not the body per se but, rather, models of the body. (Keating & Cambrosio, 2003, p. 20)

This instability is especially evident when self-tracking devices are used to provide researchers with data in standardised and aggregated form that is privileged over data that is gathered from direct observation, or put another way, data gathered through direct observation is made sense of through an evidence base (Howick, 2011, p. 4, cited in Anjum, 2016, p. 6) that is made up of aggregate data. In this context, evidence is considered to be how often an intervention works rather than if an intervention is effective. The volume of the aggregate comes from the breadth of data on a number of variables for an individual body rather than population-based aggregates where one or a few variables are considered. Evidence based medicine also assumes that there is homogeneity in a population or statistical cohort. When statistical data about individuals is sought by researchers, it is done by selecting variables that have been attributed to individuals (or individual cases). Although there are measures usually taken to account for differences between individuals by randomising samples and using large samples for test and control groups. But there is still a great deal of generalising about what interventions might or might not work in individual cases.

The way that statistical studies deal with the heterogeneity of individual bodies is to treat them as the sum of their parts which may or may not interact casually or by clustering variables together across different bodies based on biological, physiological, behavioural or social attributes and then testing correlations between variables by combining them according to some theory or hypothesis. Apprehending the complexity of sick-bodies in evidence-based medicine is achieved through modelling the multi-factorial interaction between variables that are arranged and sequenced linearly. In contrast to the paradigm of evidence-based medicine where the individual is lost to statistical representation, the approach taken in person centric, data-driven medicine attempts to reassert the individual back into information taxonomies that are likely to be more concerned with why an individual is presenting with certain symptoms. Data-driven healthcare tires to assign cause and effect relations between that can be represented and subsequently observed with the intention of tracking and assessing change over time. One of the major problems with data-driven healthcare is incisively expressed by the Australian author and physician, Karen Hitchcock who writes:

> Fundamentally, it would seem, a doctors job is to treat and protect the diseased and threatened organs –those beautiful and intricate structures that pump and squeeze and metabolise away, mostly quietly, keeping us alive … But when doctors sit in consulting rooms or walk the wards, we don't confront kidneys or hearts. We encounter complex, imperfect suffering people with specific histories, cultures and understandings of sickness and health … The medical system is constructed from tacked together fragments – GP [General Practitioner], specialists, the hospital, mental health services, community services – all of which have little contact with each other. We treat social and medical needs separately although they are entirely entwined. Hospitals themselves are designed around a body fragmented into discrete organs. Often our institutions serve clinicians, bureaucrats, and cutting-edge science better than those they are supposed to be serving. (Hitchcock, 2015, p. 2)

If by incorporating self-tracking data into existing modes of medical knowledge production, bodies might be thought of as 'walking event recorders' where 'biosensors [are used to capture] data and algorithms … process it' and this 'sets up the ability to track virtually any [health] metric' (Topol, 2011, pp. xxxx, vi–viii). But, a precautionary principle still applies. As Eric Topol notes:

> [...] as with any revolution there are important downsides to consider. Here, the concerns include the reduction of direct human contact and healing touch that may accompany increasing reliance on remote monitoring and avoidance of hospitalisations or even in-person office visits. It will be increasingly tempting for physicians to treat the virtual human being – the scan, the DNA results, the biosensor data instead of the real patient. There is a legitimate worry about the adoption of new technologies before they have been appropriately vetted and validated, or proven to be cost-effective and ideally cost saving. And certainly, the data deluge and the inability to efficiently transform massive datasets into information and knowledge loom large. An extension of the data flow issues brings us to the worry about security and privacy of digitised medical information. Ironically, the technological triumph of being able to digitise human beings creates a convergence of the real and the virtual individual, and there will be legitimate worries about depersonalisation, about treating the digital information instead of the person. (Topol, 2011)

The next section, 'Rules: The Embodied Practices of Self-Tracking' looks at this convergence of real and virtual individuals in context beyond healthcare and looks at self-tracking in the home and the workplace.

Part 2
Rules: The Embodied Practices of Self-Tracking

Chapter 3

Lessons from Electronic Monitoring

> A Fitbit is like a house arrest bracelet that punishes you for NOT going outside. (user on Reddit, r/showerthoughts, 2017)

> The bracelet itched constantly, and I was forced to quit exercising because of the sores it caused. I still have scars from wearing the shitlet. (SCRAM electronic monitor user on Reddit, 2011)

The last chapter considered the ways that self-tracking technology contributes to a disappearance of the body. I now turn to its potential to dehumanise. In this chapter I take the discussion of the contradicted nature of self-tracking technology further by viewing it through conflicted framing (Orlikowski & Gash, 1994) that arises from pre-assumed scenarios of use (Akrich, 1992) when relations between imagined devices and their realised outcomes (Krippendorff, 1989, 1990) are considered. The purpose of this chapter is to bring the quantified self into critiques of surveillance and counter surveillance (Lyon, 2001, 2013; Mann & Ferenbok, 2013; Mann, Nolan, & Wellman, 2003) and explore the potential for self-tracking technology to dehumanise users (Pedersen, 2005), biometerically profile them (Ajana, 2013), and to have lasting emotional effects (Noble & Roberts, 2016; also see Brûlé, 2015). I argue that these effects are often overlooked by sponsors of self-tracking technology and this effects designers when they are considering the scope and impact of products. This chapter seeks to advance the idea that an 'anticipatory analysis' can be developed if antecedent technologies and the flow of affordances between different classes of technology are taken into account. In comparing different kinds of technology in this manner, I argue that *anticipation* can, and should, be a methodological priority in both design, design-research, and critical inquiry into self-tracking.

The chapter proceeds by first providing a brief tracing of the evolution of electronic monitoring technology before discussing the emplacement and embodiment of electronic monitoring technology in the home and broader life-world of individuals who are subjected to house arrest and home confinement. I end by considering the affordances of electronic monitoring technology and propose a methodological approach to studying wearable technology based on the incorporation of new more intrusive technical features into existing technologies and anticipating how certain devices might evolve.

The Politics and Possibilities of Self-Tracking Technology, 65–85

doi:10.1108/978-1-80043-338-020211003

Sociologist, Mark Carrigan calls the potential use of self-tracking technology in social policy as 'coming techno fascism'. On his reflecting on his own experience with consumer self-tracking devices he writes:

> I'd got bored with the Nike Fuel Band, losing interest in the opaque 'fuel points' measurement and increasingly finding it to be an unwelcome presence on my wrist. I'd also been ever more aware of how weird my sleep patterns have become in the last couple of years, cycling between early rising and staying up late with little discernible rhyme or reason. The idea of tracking my sleep in a reasonably accurate fashion, using a degree of bodily movement as a cypher for the depth of sleep appealed to me on a reflexive level [...] [and it] really seems to work [but] consumer self-tracking devices [...] serve to normalise tracking of this sort. What comes next? How hard is it to imagine a situation where a conservative government, eager to separate the 'strivers' from the 'skivers' demand that welfare recipients submit to monitoring of their alcohol or nicotine intake. (Carrigan, 2014)

This scenario is not too hard to imagine. And in many ways consumer self-tracking technology does function like a house arrest bracelet as the Reddit user quoted in the epigraph to this chapter suggests. This chapter explores these functional similarities between the two devices and suggests that if we consider the flow of affordances between technologies designed for different applications, the potentially de-humanising outcomes of emerging self-tracking practices can be better anticipated. I argue that as key antecedents to wearable self-tracking devices, house arrest bracelets have much to teach us about the future directions of self-tracking. Looking towards electronic monitoring allows for the materiality of devices, infrastructures and environments to be taken into account without the hindrance of eudaemonia framing the technology. There are important lessons to be learned from the way electronic monitoring technologies have been adopted within the criminal justice system for anticipating problems and future directions of self-tracking technology. Electronic monitoring technology establishes unique human-to-human, device-to-device and device-to-human relations that work simultaneously to intensify and obscure the presence of technology as something that is attached onto the body.

In *Design Methods* (1970/1992), John Chris Jones noted that 'the designer must be able to predict the ultimate effects of their proposed design[s] as well as specifying the actions that are needed to bring these effects about' (n.p.). When self-tracking technology is designed, the priority for it to have compliance with a pre-determined pattern of use can overshadow other aspects of a design strategy that might result in designs being revised in ways that make the technology easier to use, more comfortable or more humane in other subtle ways. When narrowly defined patterns of use become the focus of the design-iteration process, they limit the aesthetic, ergonomic, functional and political possibilities of a design. When electronic monitoring technology is introduced into the home for the purposes of home confinement, the homespace is transformed into a place that deviates from social and cultural and personal expectations of what *home* might mean to an individual. Experiences of the home-space become conflicted and contradictory in their nature, especially when the occupant is made to wear an electronic device which dictates their daily movements, rhythms and behaviours. The experience of house arrest further complicated given that those who are incarcerated in their own homes are routinely asked to pay rents, mortgages, utilities, rates and bills in addition to paying the costs associated with administering and maintaining the technology required to sustain their electronic monitoring – especially when lived experience would suggest that electronic monitoring technology rarely functions in a way that genuinely supports this.

For example, in a patent filed in 2015 for a 'tracking device incorporating cut resistant materials', P. Christian Oliphant and Bruce Derrick note that:

> A problem with current tracking devices is that the mounting strap that is used to secure the tracking device to an individual being monitored can often be broken or cut, sometimes with only common scissors, and quickly removed. Once removed, the individual to which the device was affixed can quickly leave the area and escape monitoring and the tracking system altogether. While the complete severing of a mounting strap and the removal of an existing tracking device can often generate an alarm, the speed at which existing straps can be cut is such that the offender will typically have ample time to flee the location where the tracking device was removed before enforcement authorities can be summoned to that location. In the past, this has resulted in successful escapes by tracked individuals defeating such systems and, therefore, poses a real danger to the public. (Oliphant & Derrick, 2015)

It is due to assumptions like this that Electronic Monitoring (EM) is interesting in the study of contemporary self-tracking because its initial imagining was remarkably different from the experiences of use that emerged out of its eventual uptake in the criminal justice system. Electronic Monitoring technology blends private, pushed, imposed and exploited self-tracking and it can be used to situate self-tracking within an alternate socio-technical history. For instance, an activity tracking device and a house arrest bracelet are functionally and morphologically similar in many ways, but one of the key differences is that they are worn on different locations of the body – the ankle for the house arrest bracelet and the wrist for the activity tracker. Their underlying technical architectures and logics are similar and they have much to teach us about each other and the power relations between designers and users. Designing technologies for someone to begrudgingly live with can be a particularly fraught practice that requires broad thinking, the rationalisation of competing interests, cost-based design compromises and a range of trade-offs (Sullivan, 2016, pp. 1–8). The functionality of Electronic Monitoring depends on the combination of affordances: attachment to the body and ergonomics, location-tracking and accurate biometrics. Furthermore, the extent to which electronic monitoring technology use is sustained and complied with has much to teach us about how self-tracking technology is *lived* with.

Punitive Design

Throughout history, different forms of technology have been affixed to the body at the neck, wrist or ankle for the purposes of restraint and punishment. These are mostly for the purposes of bringing about some kind of immobile state in the wearer. Unlike other objects that might be attached to the body at these locations: pendants, wristwatches and so on, these punitive technologies are designed so that they can be locked onto the body and usually include some mechanism to prevent the user tampering with its ability to function and remain attached to the body. Even from early on in their history, cumbersome and inefficient examples of electronic monitoring technology make it easy to imagine how having an electronic monitoring device attached to the body that tracks movement might have 'unanticipated and unfortunate' (Lilly & Ball, 1987, p. 366) consequences for their users. Shackles and handcuffs, though not seen as everyday technologies can be thought of as part of the trajectory that leads to electronic monitoring. This is partly because they are considered as punitive objects where the primary function is to physically restrain the wearer. In more recent times, commercial electronic monitoring technology has been enthusiastically embraced by legislators as more economical and seemingly humane alternatives. Electronic Monitoring allows for violations to be more easily documented than in traditional methods of community-based supervision, but glitches in the technology are likely to work against the person being monitored. The intention of the original Electronic Monitoring systems was not just to enhance compliance and reduce prison overcrowding but to also help offenders gain self-esteem and be better socially adjusted (Gable, 1986; Gable & Gable, 2005, p. 21).

Over the course of their development, a number of assumptions are made about the scope and potential of electronic monitoring technology. These assumptions, over time, mask some of the negative outcomes of technically mediated home incarceration. As these assumptions about the purpose and benefits of Electronic Monitoring technology go unquestioned, they result in a narrow view of how the technology is applied, managed, regulated and studied. As one commenter on one of the many videos showing the various ways that an Electronic Monitoring device might be removed without setting off a tampering alarm suggests, 'there should be wrist monitors that look like [an] Apple iWatch instead of ankle bracelets'. Script analysis (Akrich, 1992; Latour, 1992; Verbeek, 2006, p. 31) offers a way to ratify the mode of thought referred to as 'design intuition' and explain why this is unlikely to happen (Wendt, 2015). This intuition places emphasis on the possible but not necessarily realised juxtapositions of bodies, environments, technical objects and knowledge systems of reformative or punitive justice that underlie a device's technical logic. As just one element of the different 'actionable combinations' (Letiche & Lissack, 2009, p. 62) that make up electronic monitoring systems, tamper proofing, a feature that is predicated on certain assumptions about pre-existing patterns and motivations assumed to be part of a user's life. As Akrich writes:

> Designers thus define actors with specific tastes, competencies, motives, aspirations, political prejudices, and the rest, and they assume that morality, technology science, and economy will evolve in particular ways. A large part of the work of innovators is that of 'inscribing' this vision of (or prediction about) the world in the technical content of the new object [...] the end product of this work [is] a 'script' or 'scenario'. (as cited in Fallan, 2008, p. 63)

This means that at the outset of the design process it is likely to be assumed that just by virtue of the circumstances necessitating the use of electronic monitoring devices that the user will be inclined to want to tamper with it. Script analysis suggests that technical objects prescribe human interactions with it through the characteristics that are inscribed into it during its imagining, iteration and manufacture. For instance, a cup that is made of paper suggests disposability whereas one made of fine porcelain suggests fragility and value; a device worn on the wrist suggests active and productive purposes of use, whereas one that is worn on the ankle suggests punitive ones. In the context of self-monitoring technology, script analysis offers a way to document and understand a 'designer's more or less informed presumptions, visions or predictions about the relations between [a technical object] and the human actors surrounding it' and potentially establish this perspective outside narrow definitions of user behaviour where the fact that in certain contexts some users will want to ignore, discard and reject elements of the standard account for how a technology is used and what purposes it has been optimised for might be overlooked (Fallan, 2008, p. 63). But at a deeper level, script analysis offers a nuanced vocabulary for understanding the political stakes linked to the forces guiding the development of embodied self-monitoring technology. First, the notion of de-scripting refers to a mode of critical analysis that attempts to read the motivations, aspirations, political prejudices embedded in a given technical artefact through the object itself, if it is available, and through material and documented traces of its design and development and people's experiences of its use. Affordances then are something that are the result of pre-scribed intentions for the technology or derived by subverting the intended logic of the device by going off script and de-inscribing the technology. Re-inscription is when new or alternative affordances are absorbed back into a design (Fallan, 2008, p. 64).

The ways that design-guiding discourse flows between the sponsor, designer and user of self-tracking technology is a complex configuration of interpretative choices that are made when problems, use cases, prototypes for a technology are formulated in a collaborative manner though assumptions about what minimum standards for a design should be and what acceptable conditions of use are prevail against commonsense. When users are consulted in the design process, or when they are included in observational or ethnographic research into the use of electronic monitoring technology, the extent to which problems (and ways to cope with them), particularly admitting non-compliance which has implications for users, are meaningfully dealt with is likely to be partial at best.

An affordance, then cannot simply be 'built into' or 'read out of' (Fisher, 2004, p. 26) a technical object functioning in an average scenario of use. Such assignments of an affording capacity for technology are made meaningful not only by corporeal, observed materialities, but also through semantically and experientially derived frames of reference that over time, and through processes of diffusion and assimilation, become attached to the prototyped and developed examples of a device. The experience of an electronic monitoring system, when considered through its inscription, highlights the

> …symbolic qualities of man-made forms on the [cognitive, gestural and social] context of [how electronic monitoring devices are used] and the application of this use to [the professional practice of] industrial design. (Krippendorff & Butter, 1984, as cited in You & Chen, 2007, p. 26)

The inscription of technical logics translate to an application of knowledge about modalities of use that relates symbolic qualities to user compliance with design motifs which are materialised and accumulated as 'features' as a class of technology evolves.

The Evolution of Electronic Monitoring

Wearable electronic monitoring for home confinement was first developed as a legal and judicial measure to restrict an individual to a physical location or dwelling for punitive purposes (Lilly & Ball, 1987). As an alternative to imprisonment in jails, house arrest was enabled by the use of electronic monitoring equipment which were considered to be more cost effective and lenient alternative to housing low to moderate risk offenders in correctional facilities with other prisoners. The use of electronic monitoring, even from its early applications was not limited to penology and was also used for monitoring political dissidents, protecting the children of high-profile individuals from kidnappings and for remote patient monitoring in healthcare.

The first experimental uses of electronic monitoring began in the 1960s when Robert and Ralph Kirkland Schwitzgebel (later Gable) began designing portable transceivers capable of recording location data that were given to volunteers who had criminal records (Schwitzgebel, Schwitzgebel, Pahnke, & Hurd, 1964; Schwitzgebel, 1972). By 1965, Radio Frequency (RF) transmitters were being prototyped for monitoring prisoners. Examples of this early generation of the technology typically comprised of a transmitter that could be worn on the body that was powered by battery packs attached at the waist on a belt. The wearer's location could be determined by an array of multidirectional antennas (Schwitzgebel & Schwitzgebel, 1974). The Schwitzgebel brothers had been influenced by psychologist, B.F. Skinner. Their prototype was named the 'Behaviour Transmitter-Reinforcer'. It was capable of sending two-way messages between a user and a base station about the users activities and emotions while in natural social environments. The primary goal was to provide feedback to the offenders for moral support during their efforts to be become rehabilitated. The inspiration for the device came when one of the brothers, Ralph Kirkland saw the film *West Side Story* particularly the scene in which the film's hero is killed by an opposing gang member. Schwitzgebel thought that if the hero could have received a warning then his life might have been saved. Skinner's work aimed to explore how social behaviour might be influenced by the specific nature and timing of consequence.

Although the brothers Gable attempted to explore a range of potential uses for the technology they were 'rebuffed by virtually every professional organisation, foundation and citizen group to which [they] turned' (Tribe, 1973, pp. 331–332, as cited in Burrell & Gable, 2008, p. 103). A range of factors contributed to the lack of interest in developing what, thanks to the editors of *Harvard Law Review* became known as the 'Schwitzgeble Machine' (Fox, 1987). The first would have had something to do with misrepresentation of how the technology worked. All sorts of myths had begun to circulate and grow over time, for instance one was that the device used brain implants to transmit information into, and out of, the users' minds for instance (Everything2.com, 2001). Beyond this, there were, understandably, issues with the social acceptance of encouraging behavioural change by using positive reinforcement and at the time there was not enough public knowledge about Skinner or his work for the device not to be overshadowed by these negative associations.

It is also not difficult to understand how people would have associated the technology to a real-world version of Big Brother from George Orwell's dystopian novel 1984. Finally, the device was functionally and philosophically a radical departure from wearable correctional technologies being used at the time and considered by many to be just too much of a departure from what already existed. But a decade later, single assigned radio frequencies were being experimented with in clinical settings where telemetric monitoring systems with electrocardiogram sensors were being tested on patients to be able to remotely monitor body functions (Depedro & Cannon, 1975; Mandel, Ong, & Singer, 1975). The applications for this type of technology were limited because in the 1970s telemetric and telephonic technology was not capable of coupling location data and information on vital signs in a single transmission.

The 1970s saw significant correctional policy reform, particularly in terms of being more punitive and offence-focused and less concerned with offender rehabilitation (Cullen & Gendreau, 2000). With the shift away from rehabilitation and towards surveillance and the enforcement of strict conditions over offenders' lives, the prison system in the US drifted into crisis and this made it more amenable to taking unorthodox proposals for technology more seriously – especially given that microprocessor technology was becoming more common and less costly. The generic term 'house arrest' gained currency after the passage of the US Correctional reform act in 1984. Shortly after, an electronic bracelet was used for the first time to monitor a prisoner's compliance with a curfew. The chief sponsor for this technology, New Mexico district judge, Jack Love had been inspired by two things he saw in a newspaper in 1977. The first was a Spiderman comic where a villain clamped a bracelet style radar device on Spiderman to track his movements. The second was a device called a heifer, a radio transmitter that is implanted under the skin of cattle and a handheld detector that received information about the animal's history and diet requirements (Holm et al., 1977, as cited in Burrell & Gable, 2008, p. 104).

By the 1980s systems that comprised of assigned passive tags to moveable objects that could be interrogated by fixed active transceivers in the immediate environment were being prototyped. Full blown functional electronic monitoring systems did not arise until there had been sufficient advances in the areas of database architecture, interface design, mobile telephony and materials technology. Love had previously served as a Federal public defender and had sought ways to keep non-violent offenders out of overcrowding jails. Love approached a few technology companies looking to develop a prototype and commercialise the idea. Most of them were not interested except for Michael Goss, a sales representative at Honeywell. In 1982 Goss left his job at Honeywell and with a 10,000-dollar bank loan established a company called National Incarceration Monitoring and Control Services (NIMCOS). Goss built several transmitters that were about the size of a fist that were designed to be strapped to an ankle. The device sent an RF signal every minute and this signal could be picked up by a receiver connected to a telephone line and transferred to a mainframe computer. It was a crude telemetric system with a range of about 45 meters and the fact that loss of signal would occur at distances than this was used to detect non-compliance with home detention.

In 1983, three people sentenced by Judge Love were fitted with the device as a means of enforcing probation curfews (Burrell & Gable, 2008, p. 105). The first was a person with an infant child who had a steady job. They had been convicted for writing fraudulent cheques. Although this probationer successfully completed a 30-day period on a curfew monitored by the device, they reoffended and were arrested for shoplifting. The second, a Vietnam War veteran studying to be a computer technician was placed on a 30-day monitored curfew for violating his probation by receiving stolen goods. Although this probationer did not violate the conditions of the curfew, they were sent to jail after five days after returning to the local detention center where they were being housed intoxicated (Burrell & Gable, 2008, p. 105). The third was an individual who was convicted for a second Driving Under the Influence (DUI) offence successfully completed 30-days of monitoring (Burrell & Gable, 2008, p. 105).

Although Goss's experimental trial of Electronic Monitoring was successful in terms of how the technology performed, the GOSS link as it was called ceased operations because Goss had run out of money. The device though functionally effective did not demonstrate a reduction on recidivism and ultimately, Goss, having not forgotten about Love's inspiration for the device coming from cattle tagging, offered the design to Boulder Industries, a company that were also struggling to market their own electronic dairy identification system. Boulder Industries CEO, David Hunter saw an opportunity. He loaned NIMCOS $250,000 USD and arranged for a company called Control Data (Minneapolis) to market the technology. The arrangement failed on account of financial problems at Control Data and Boulder Industries eventually purchased the assets and commercial rights for NIMCOS for an additional $250,000 (Burrell & Gable, 2008, p. 106).

At the same time, Judge, Allison DeFoor of Monroe County Florida sentenced a 28-year old carpenter to a weekend of home detention for driving without a licence (Shillington, 1983, cited in Burrell & Gable, 2008, p. 106). Thomas Moody an inventor who had a company called Controlled Activities

Corporation (CONTRAC) manufactured and installed a device which consisted of a wearable radio transmitter that was housed in a five by three inch plastic cylinder that could be attached to the probationer's ankle with a strap, the other component was a receiver that was installed into the dwelling where the probationer was to be confined. The anklet sent signals to the base-station every 35 seconds that was linked by telephone to a computer located at Moody's monitoring service, Omni Communications, Inc. In 1984, a larger experimental study commenced between CONTRAC, the Palm Beach County's Sherriff's Department and Pride Incorporated, a not-for-profit organisation that runs monitoring programs. The first probationer a 20-year old student convicted of a DUI was confined to their parents' home on nights and weekends. The probationer was required to pay $250 USD to participate in the trial.

Electronic Monitoring systems of this era relied on the same type of components that were used in emergency aircraft beacons and lacked the capacity to be manually activated by the wearer. With the introduction of cellular telephones and the Global Positioning System (GPS), combined biometric and location data could be transmitted between devices that were attached to the body, cellular phone towers and remote computers. This development enabled the wearer's location to be assigned to other data points relating to the wearer's physiology or mobility-based patterns of behaviour. Since the late 1990s when the GPS was declared a dual use system accessible for civilian purposes, there has been a steady development of location-based and biometric electronic monitoring technology.

Throughout the 1980s and 1990s both the number of cases and manufactures increased (see Renzema & Mayo-Wilson, 2005) though the electronic monitoring industry itself was quite volatile and subject to growth, shrinkage and consolidation. The increase in adoption of electronic monitoring during this time was also in part facilitated by the broader enthusiasm for technological solutions across multiple sectors, marked by: (1) the popularisation of personal computing and portable electronic lifestyle technologies such as walkie talkies (two way radios), handheld computer games and the Walkman (portable cassette players); (2) using information and computing technologies to automate and make more efficient management and administration processes in office environments.

In a report to the US Department of Justice entitled, *Preventing Crime: What Works, What Doesn't and What's Promising*, Sherman et al. (1997) list electronic monitoring as something that 'doesn't work' when compared to the 'placement of similar [low risk] offenders under standard community supervision without electronic monitoring' in terms of its reduction in reoffending (1997, p. 9; see Austin & Hardyman, 1991; Baumer & Mendelsohn, 1991). The Secured Continuous Remote Alcohol Monitor (SCRAM) was introduced into the market for offender management technology in the early 2000s. Continuous monitoring was a solution to the issue of random testing which was only partially effective and readily subverted in terms of compliance. There are an estimated 483,000 persons using a SCRAM device worldwide (Alessi, Barnett, & Petry, 2017, p. 417). The SCRAM device detects the presence of alcohol vapour near the skin by way of an electrochemical sensor. It has two circumvention detection sensors which monitor skin contact and temperature. The device collects a sample every 30 minutes under standard operating conditions and this automatically becomes more frequent if alcohol is detected, or, if the device detects a tamper attempt. Date and time stamped data are uploaded to a central server via a USB connection or modem.

Doing Time in the Home-Space

To understand the impact of electronic monitoring devices is to also consider how the spatial, temporal, public, and private aspects of home-dwelling are arrayed and reconfigured when an electronic monitoring device is attached to a person serving their sentence out in the community. Pre-existing attitudes and expectations about prison rights are one of the main drivers of the technological developments being made in house arrest technology and it is the assumptions made in this regard that sets the agenda for new technical trajectories that lead to the incorporation of certain features into existing systems. For example, tamper proofing, sampling of biological material though the skin, the incorporation of biometric sensors, the integration of other types of stored data about users and the use of algorithms to predict certain behavioural or physiological patterns in the user are all likely to be given greater priority than optimising the technology to improve user health and wellbeing. The notion of market responsiveness (Hurley & Hult, 1998) dictates that technology innovators anticipate the needs of potential or current users and understand problems through the requirements of intermediaries rather than those most affected by house arrest –the individual wearing the device.

When an individual is placed under home incarceration or confinement, the home-space is transformed in many ways. To give these transformations ontological grounding, one can consider the layers of meaning ascribed to the home-cum-prison that would not be apparent to the casual observer. For those being monitored in their own homes, the prisonised home-space juxtaposes several places and at the same time. It becomes a space of respite, but it is also transformed into a space of structured discipline and rehabilitation when one of its inhabitants is required to embody abide by the logics and flows of the monitoring device. The home-cum-prison only fully functions in this regard when there is a break with normal experiences of time and daily routines associated with life on the outside. Wearing an electronic monitor alters not only the nature of the hours and days spent at home but also their relation to the past, present and future as a function of duration. The use of in-home electronic monitoring is predicated on a system of opening and closing the home in ways that both isolates the occupant from certain people, social interactions and mobilities and at the same time makes the home-space penetrable to non-negotiable entry, both physical and virtual, by law enforcement agents or their delegates.

The individual, the technology and the embodied subjectivity resulting from the introduction of a technological an ongoing presence into the individual's lifeworld make the standardised judgment of what an average user's experience might be difficult. Among the perceived benefits of Electronic Monitoring is the idea that house arrest is – according to proponents – a less coercive measure given that an individual opts in to wearing an electronic monitoring device as an alternative to serving time in prison. As a form of imprisonment, house arrest is rarely imposed as the sole available option by a judge or prosecutors. Moreover, it is thought that house arrest grants an individual the right to work and protects individual liberty and intimacy compared to the communal nature of prison-life and social conventions that displaces an individual from the positive virtues of having a private home-space and access to a community and its resources. House arrest programs also promote forms of public humiliation or stigma when electronic monitoring devices are visible to others.

Responding to the question: 'what is it like to be on house arrest' on Quora, an online platform for crowdsourcing answers to various questions on a variety of topics, someone with the moniker, 'John Wayne' makes the distinction between home incarceration and house arrest:

> Home confinement: I had a[n] 8p.m. curfew, ankle bracelet and a separate monitor that required a phone line. The ankle bracelet was about half the size of a pack of cigarettes. It required a battery which lasted 8 months or so. It did not have GPS or whatever else. If the monitor did not sense the bracelet within the proximity by 8p.m. it would send a[n] alarm to my [Prison Officer] PO. It required a phone line and $40 service fee to BI which is the company that monitors and provides the equipment. I only missed my curfew once and nothing happened. If I missed it again, he threatened to lock me down for 24 hours. After the first month curfew was still 8p.m. but I was gradually able to request it be extended to 9-11 depending. I adjusted to it rather quickly and it wasn't too bad. Worse part was the stress of my PO randomly showing up. Whether I was doing anything wrong or not it was still stressful. Obviously, you miss being able to enjoy night life, late night errands etc.
>
> Home incarceration: The biggest difference between home confinement and home incarceration is that under home incarceration you are locked down 24/7. On rare occasion I could leave for court, lawyer meetings and emergencies. Besides that, I couldn't go anywhere. No holidays, haircuts, groceries or whatever else. I wore a much larger bracelet which was probably about the size of 2 packs of cigarettes stacked one on top of the other. This bracelet had to be charged nightly. It was equipped with cell phone and GPS and did not require a phone line. It was also [two]-way. My PO could page me which would cause it to start beeping. I would press a button and it would play a pre-recorded message such as 'call your officer now'. My PO said that he would allow me to return to home confinement after 60 days but somehow, he made excuses and it ended up being close to 100.

> Even when I was allowed to return to home confinement it was still much more limited. Basically, work and a small amount of time allotted on weekends for whatever else. The first week was pretty easy. After that it was pretty rough for the first month. You eventually get used to it. Under home incarceration a phone line was not required but the monthly monitor fee was $100. (Quora, 2016)

In contemporary terms, Electronic Monitoring serves as 'a socially expedient intermediate sanction that is more punitive than traditional probation, but less harsh than incarceration' (Gable & Gable, 2005, p. 2) but the extent to which it might be deemed less punitive is contingent on intersecting factors associated to risk, race, gender, economic status, judicial environment and sociocultural factors. Despite their non-physical tethering effect, electronic monitoring devices are embodied differently when the wearer is under home incarceration or home confinement. For example, concerns about the extent to which the technology is discrete is different in the privacy of one's home if they live alone, with family or with others depending on the circumstances. The same can be said when the user is in various public spaces where there may be more (or less) stigma associated to a visible electronic monitoring device.

In a Reddit AMA session entitled, 'I am a person who monitors and track[s] people that are on court ordered ankle monitors. Ask me (almost) anything' a user asks 'Do you watch them real time? Do you and the gang at the office ever take bets on their destination? Is it exciting when they make a break for it, or do you dread all the extra paperwork you'll have to do?' To which the original poster, responds:

> They are watched close to real time. There is a delay with the devices which can be delayed for up to 10 minutes. We can activate the device to a more rapid mode that will keep constant communication so we can see real time data, however, we generally will only do this in emergency situations. It is not entirely too exciting when they make a break for it. I just want to make sure that the device is returned to me. If you are on a monitor and you want to make a run for it, please just cut the monitor and leave it somewhere safe. Do not make it harder on yourself by destroying the monitor, you will be charged with extra penalties and fines if you do and it is not cheap.

Punitive aspects of Electronic Monitoring are not only the result of 'more rigorously enforced compliance' (Gable & Gable, 2005, p. 21), they are also the result of a complex array of psycho-ergonomic factors that result out of: (1) how the technology is designed and functions and (2) how easily (or not so easily) the technology can be lived with and integrated into its users lives with all the punitive measures of both technological and legislative compliance in place. One of the problems with the way we might think about how Electronic Monitoring technology, and by extension self-monitoring technology more broadly might be made more efficacious is that most of the rigorous study into what could be done to improve systems is centred around the measurement of progress towards the objective of reducing recidivism and reoffending. For example, in a meta-analysis of 381 articles on the effectiveness of Electronic Monitoring, Renzema and Mayo-Wilson note:

> It is hardly surprising that recidivism has not been reliable reduced by an intervention that is typically quite short, applied in a standard fashion, and applied to a diverse group of offenders for whom it may or may not have any relevance to their motives for offending. Extant Electronic Monitoring programs seem akin to giving Aspirin to a mixed group of hospital patients and then wondering why their underlying diseases have not been cured. (Renzema & Mayo-Wilson, 2005, as cited in Gable & Gable, 2005, p. 22)

But Electronic Monitoring appeals broadly to correctional reform agendas based on economic rationales with a veneer of humanitarianism. According to the PEW Charitable Trusts (2016), in the United States alone more than 125,000 people were using electronic monitors compared to 53,000 a decade earlier. One of the reasons that the electronic monitoring devices are so cost-effective is that, problematically, the financial burden associated to operating and maintaining the device is passed onto the user, approximately $10 USD per day after an initial fee. The day-to-day monitoring is administered by local justice organisations or outsourced to third party intermediaries who deal directly with the user. The user pays business model has a range of problems particularly in that one of the most championed benefits of electronic monitoring is the idea that those being monitored can obtain and hold down jobs, helping them pay for the device and reintegrate into the community. For example, Maya Schenwar, a prominent writer on the topic of humanitarian issues associated with prisons points out that in both of these instances, a visible ankle monitor acts as a 'scarlet letter [for] those permitted to go to work [because they still have] a difficult time finding or holding down jobs' which is a problem given that 'gaining employment [in the first place] is a crucial step in community based rehabilitation programs' (Schenwar, 2014; Schenwar & Jan 2015).

In a *New York Times* (2019) article, Ava Kofman reports on a 19 year old who had been using an electronic monitor. On this person's experience, Kofman writes:

> [He] signed a contract stating that he would charge his monitor for an hour and half each day and 'report' to Emass with $70 USD each week. He could shower but was not to bathe or swim (the monitor is water resistant but not waterproof). Interfering with the monitor's functioning was a felony. [He] assumed that the GPS supervision would be a minor annoyance. Instead, it was a constant burden. The box was bulky and the size of a fist so, so he couldn't hide it under his jeans. Whenever he left the house, people stared. There were snide comments ('nice bracelet') and

> cutting jokes … The biggest problem was finding work … To help improve his chances, he enrolled in [a job training class] for people re-entering society. One afternoon … he and a classmate role-played how to talk to potential employers about criminal charges. [He] didn't know how much detail to go into. Should he tell interviewers that he was bringing his pregnant girlfriend some snacks when he was pulled over? He still isn't sure, because a police officer came looking for him midway through the class. The battery on his monitor had died. The officer sent him home, and [he] missed the rest of the lesson. With all of the restrictions and rules, keeping a job on a monitor can be as difficult as finding one. The hours for weekly check-ins at the downtown Emass office – 1 p.m. to 6 p.m. on Tuesdays and Wednesdays, and 1 p.m. until 5 p.m. on Mondays – are inconvenient for those who work.

The cost of monitoring and maintaining electronic monitoring devices are normally covered by the user from their own funds and faults or failures of the technology may result in inappropriate questions about tampering and user compliance. When undergoing home incarceration, prisoners face further, more basic, challenges that their counterparts in prison do not. Daily meal planning with nutritional and caloric management and structured outdoor exercise time are available albeit sub-standard in terms of quality of food and facilities (Kilgore, 2014) that are not available for those incarcerated at home, though such features could be integrated into house arrest systems if one looks towards the consumer health and fitness tracking technology market for inspiration.

The typical duration of Electronic Monitoring is about a year plus some degree of post monitoring after care or counselling (Cullen & Gendraeu, 1989). Gable and Gable note that 'a monitoring transmitter might be conceptualised as a 'social prosthetic device' similar to a walker that is downgraded to a crutch, then to a cane, and finally abandoned' (2005, p. 24). The psychological toll of perpetual surveillance, especially feeling trapped and as if they were service a sentence before even having gone to trial. The person referred to in Kofman's article, for instance:

> […] was never really sure about what he could or couldn't do under supervision, … when his girlfriend had their daughter, … [He]left the hospital shortly after the birth, under the impression that he had a midnight curfew. Later that night, he let his monitor die so that he could sneak back before sunrise to see the baby again. (Kofman, 2019)

Inscribing Compliance

The development of new generation Electronic Monitoring technology should be viewed as open, replaceable, and amenable to refinement and improvement by incorporating features being developed in other applications of similar technology. The inspiration for new forms of Electronic Monitoring should come from science fiction and popular culture as it did for Love but also from people who have intimate experience of using the technology across a range of different conditions.

Re-framing self-tracking technology with an eye to the history of Electronic Monitoring devices as technical antecedents offers a predictor for the influence that the industries forming around personal analytics might have, especially in terms of how the surveillant and punitive affordances of self-tracking are enveloped as a 'rhetorical gambit' that obscures certain 'pre-existing [and stigma-producing] social formations' (Winston, 1998, p. 2). As Winston notes, the repurposing of technology:

> ...can be seen across [a] diverse range of technologies and [across decades] of development and diffusion. Consider, for example, the ways in which ideas for devices occur, the importance of science, general knowhow, the relationship of prototypes to 'inventions' and the balance of forces pushing and inhibiting technologies. A model to reflect these patterns implicitly suggests the primacy of the social sphere as the site of these activities, conditioning and determining technological developments. It allows us to go beyond a straightforward account of technological history to pose a more general questions about how the pattern of innovation and diffusion [...] illuminates the broader role played by such technologies in our civilisation. (Winston, 1998, p. 2)

Repurposed technologies 'perform' the work of bringing together the outcomes of different sector-specific R&D of the day, reflecting the constraints and norms of a particular era and often conflicted nature of purpose arising from this lateral and combinatorial approach to design. Therefore, the development of a particular device will be structured by the research that has gone into the, design of 'new' technologies as being directly proportional to the investment made in unearthing technological innovation in universities, corporate labs or maker communities. For Winston, what moves technology from R&D to the prototype stage is ideation which functions to attract funding and generate interest among technologists, administrators, developers or solution-seekers (1998, p. 4) into the process of design. This occurs when a device is envisaged as having an application as the preferred solution to a particular problem and all involved parties have shared and individual interests that could be met by the development of the technology. The envisioning of these applications, as per the model, are defined and constrained by forces operating in the social field that surrounds the transition from technical competency to technical solution. This introduces the question of why, then, are some prototypes more amenable to being taken up and developed and others less so? Why do certain market-ready technologies come into being through the work of different groups working in isolation from one another? How is an affordance of a particular component of a technical system considered to be appropriate for a particular use case relative to the drivers which demand it?

As Winston notes, '[t]hey can range from the objective requirements of changed social [norms] … through to the subjective whims of perceived needs' (1998, p. 6). Certain types of design enhancements can be rejected because the supervening social necessity is not yet articulated in a clearly defined set of norms. Conversely, it can be accepted when a supervening social necessity has created a partially articulated need which is then given gravity against some or another social or political driver. The proliferation of prototypes by virtue of these forces may occur in parallel, differentiated by their use-case which might be in different domains or problem-spaces. Partial realisation of the capacity of prototypes may occur when prototypes are picked up and applied to areas where they are do not perform optimally but by the time this become apparent technical 'lock in" has set in which limits the ability of technologists and technology vendors to rethink the fundamental design of a given device. Given that the purpose of Electronic Monitoring devices is to constrain their users whilst granting conditional personal mobility, an ontology of space and time is embedded into the logic of these devices which suggests to designers that user compliance is something that should be understood through data capture and the continuity of information transmission. Because the home is also a space for the expression of private desires, difference from the outside world and a safe space creative experimentation, people under house arrest might develop unique workarounds so that they can act with some degree of strategic agency and non-compliance in terms of how they live with their monitoring device.

That 'people will strive to develop competencies with new technical objects in order to feel comfortable in their presence' (Krippendorff, 1989) is something that is commonly assumed in the design of human-centred technology and we should also assume that this would also include ways to interfere with the normal operations of the device. In a study of non-compliance and house arrest, Hucklesby found that technical issues with the ankle bracelets were often exploited to engage in non-compliant behaviour without consequence by simply 'blaming the equipment' (2009, p. 263). A Reddit user posting during an Ask Me Anything session (AMA) about what it is like to be under house arrest explains:

> I was working for a computer repair shop at the time and the techs and I kind of dissected the thing, did a lot of research on it online, and did some experiments (under the guise of 'I was at work so it doesn't count') and we determined that the damn thing wasn't even on. Turns out that the particular one I had [came] with a service with 2 monitoring options (agents can pull reports as needed or it is monitored for activity 24/7 and the agent is called anytime the person goes anywhere) We figured that the supervising agency had to pay for any time the bracelets were activated. Based on a combination of what we found out online, the lights on the bracelet when I was in various locations, and the reaction (or non-reaction) of my agent to my movements (like one time I stayed at work until [midnight] when I was supposed to leave at 5p.m.) we figured it wasn't even being monitored. We were right. From that point on I just did whatever I felt like.

Compliance in this sense conflates rules associated to the physical device itself, its software, a prisoner's sentence, and the law more generally. Errors in compliance 'largely arise out of a mismatch [between] what a [device] can actually do and what it symbolises to [sponsors and users] as being capable [or good at] doing' (Krippendorff, 1989, p. 8).

Two theories of compliance can be applied here to better understand its relationship to the individual approaches to and resistances of conditional agency during house arrest. These are formal and substantive compliance. Formal compliance refers to the minimal requirements required to meet the terms dictated by the device, the sentence and the law. Substantive compliance refers to deep engagement with the terms of the sentence and deep engagement with the overall ethos of house arrest at the level of ideology (Robinson & McNeill, 2008, p. 434). When a feature like tamper-proofing is conceived of there are limited options available to deter how removal is prevented and how a removal event is communicated to a monitoring agency. One could argue then that the relationship between the material and semantic aspects of tamper-proof devices can be framed by the shared presumption that an individual who is under house arrest will not entirely perceive their conditions of their incarceration through the full set of experiences that the technology could create for someone but focus on those which are considered subjectively by that user to be frustrating.

Understandings of tamper-proofing then are most readily observed through the behavioural actions and attitudes (or documentation of these actions and attitudes) of someone who is under house arrest and wearing an ankle monitor. These interactions with the device may reflect aesthetic concerns or ergonomic frustrations, or both – for example, the device being visible or awkward looking when certain types of clothing are worn or uncomfortable to wear while sleeping. The socio-technical script comprises in equal parts the physically afforded cumbersome aspects of wearing the device, their experiential embodiment in the form of frustration and discomfort and the extent to which a wearer develops tolerance or circumvention strategies. The domain extension of features that are inscribed into Electronic Monitoring technology into other intimate areas of life means that established, emerging, or potential markets are treated as outcomes of processes that contribute to the diffusion of technical systems in which wearable data-garnering devices are already present. These processes are then shaped by the interests of those who need to cater to a market and then by extension, contribute to the shape that market takes as it forms and approaches maturity. The appropriation of technology designed for the criminal justice market might seldom yield radical innovation in the form of spin-offs because their logics and architectures do not necessarily cater to other marginal markets and niches that are not necessarily profitable if those developing the technology have never dealt in these markets before. Rather, the tendency within Electronic Monitoring technology might be for incremental innovation through existing parallel lineages of technical affordance, which are assumed to be relevant during the R&D process.

The challenge of developing a critical understanding of self-tracking technology lies in how the social dimension of the technology, which naturally changes over time, is reasserted against material understandings of the technology. Anticipation as a methodological approach seeks to extrapolate from a known understanding of things, a deeper understanding of hidden surfaces, relationships and interdependencies. In anticipating the socio-material evolution of electronic monitoring use, we may be able to consider some of the iatrogenic outcomes of technical developments in house arrest systems so that trade-offs can be ethically and pragmatically assessed relative to the uncritical acceptance of tighter mechanisms of compliance and control, as simply a matter of 'human contingency' (Akrich, 1992, p. 205) as a concern in a design brief. What anticipation offers is a nimble discourse that extends the 'finite vocabulary' (Krippendorff, 1989, p. 10) attached to house arrest as a scenario of use for electronic monitoring systems. As Krippendorff argues:

> Only when designers [and researchers] are able to see themselves as part of the larger system of meaning, an ecology that guides the creation and use of artefacts, changes in the material world including their own understanding of it, can designers assume responsibilities for their own ecological interventions, which ultimately are interventions into [...] practices of living. (Krippendorff, 1989, p. 31)

For more humane Electronic Monitoring technology to become a feasible option it would have to have: (1) clearly articulated principles of social learning, motivation and self-efficacy theory translated into functional technical scripts that could be embedded into the logic of an electronic monitoring or self-tracking system and have balanced positive and punitive affordances for the user; (2) changes to the procedures of third parties incentivising the collection of data, acting on data transmitted from devices both in terms of the permissions for access and sharing, and also the extent to which inferences about human behaviour and learning can be made by third parties based on the data; (3) the identification and active but incremental phasing out or redevelopment of affordances that contribute to adverse effects (Burrell & Gable, 2008, pp. 112–113). As Burrell and Gable note:

> Positive [reinforcement] could be delivered in conjunction with alcohol monitoring systems [such as] SCRAM that tests perspiration excreted through the skin. Because the monitoring agency can set the interval at which the bracelet does the testing, a detailed pattern of alcohol use or abstinence can be obtained. Other remote alcohol testing devices that use a standard telephone line [...] are particularly amenable to re-design to the extent that they could permit immediate positive feedback at remote locations for clean tests and other desired behaviour [... or] a network of transceivers might be periodically installed and moved in order to create an 'electronically enriched' urban area where offenders could become eligible for a reward if they spontaneously exhibited cooperative rather than aggressive social behaviour. (Burrell & Gable, 2008, p. 114)

Table 3.1 shows the affordances and effects of devices when they become part of data garnering systems.

Table 3.1. Affordances of Data Garnering Electronic Monitoring Systems.

Affordance	Effect
Active/passive data capture	Can capture data on which inferences can be drawn.
Adaptive/contextual data capture	Can record contextual data (physical environment or of an individual's location and movements) and extrapolate information about events in time.
Addressability	Has at any moment a unique identifier that can be potentially knowable such as an IP address, MAC address or GPS coordinates.
Associability with living beings	May have degrees of personal association (physical, emotional, legal), may be customisable in order to meet user preferences regarding ergonomics and aesthetics.
Autonomy	May be fully autonomous or have some degree of autonomy from human users or systems of which they form part, their autonomy may have varying degrees of sophistication.
Dependency	May depend on remote services or external data sources or technical infrastructure.
Location awareness	A sensor-object or system of such objects can be locatable in physical space and the object itself can record its physical location via the GPS at a fairly granular level.
Human computer interaction platform	May mediate some human function be being worn, embedded or implanted into the body.
Identifiability	The object may have one or more unique identifiers that are potentially knowable to eternal parties.
Can be located	A single device and/or those devices it communicates data with can be found within a sociotechnical network.
Mobility, location worn on body	The object may be operational in public and private space and collects data as users move relatively autonomously.
Operational, economic and social	Features of the object may be beneficial to some parties and detrimental to others.
Prevalence	The object can appear in many places (pervasive) or it may appear at 'all' relevant places (ubiquitous).

Table 3.1. *(Continued)*

Affordance	Effect
Pattern	The context and frequency with which a user uses the object and under what circumstance.
Volatility	The object may have variable levels of connectivity, security or other elements of its design might influence its data processing capabilities.
Vulnerability	The object may more or less vulnerable to security breaches, theft, and physical damage or destruction or discontinuation of use.

Source: Adapted from Manwaring and Clarke (2015, p. 600).

A problem that needs to be pointed out with the types of frames that are established when 'product semantics' are conveyed in Electronic Monitoring technology is that when they are part of a conversation between designers and users (see Norman, 1999, 2008), strategic human agency is presented as being contingent on design choices and in the case of electronic monitoring devices this agency (and comfort) is a conditional and changeable state. Such a design strategy assumes that the material properties of a device, when communicated in the context of utility, plays a role in the communication of how meaningful a technical object is and prescribes the things that people might achieve by using it. This scripting carries a different materiality into the final product, one not so concerned with the physical manipulation of an object or the movement of that object between use contexts but rather, with controlling what that object could mean in the minds of idealised users as one gives oneself over to the logic of the device, the system and the frame which surrounds them both.

This type of speculative modelling could link different configurations of the same set of human-technology relations based on the difference between the assumptions of the discipline in which the analysis is occurring, and the objectives of the system under examination. This helps us grasp the material traces of dynamic design processes, see change over time, speculate on the rationale of decisions and identify mechanisms of marginalisation, and overall access an emerging dimensional politics flowing though this class of technology. To access the deeper confines and intimacies of the everyday lives of electronically monitored persons, especially given that we know little about how individuals adjust after serving their sentence, requires a two-fold examination of the afterlives of such technologies.

This brings us to the question of how this might be achieved especially if we are seeking to understand the dehumanising characteristics of electronic monitoring technology given that off script use, non-compliance and workarounds are a key part of how these understandings are generated and the appeal of Electronic Monitoring technology should be tempered by a pragmatic concern for sound analysis of user experiences to guide its use (Burrell & Gable, 2008, p. 115) along with an empathetic approach to its scenarios of use and how experiences in one scenario might apply to others.

Chapter 4

Sociometry

> For a long time, only one area of human activity appeared to be immune. In the cozy confines of personal life, we rarely used the power of numbers. The techniques of analysis that had proved so effective were left behind at the office at the end of the day and picked up again the next morning. (Wolf, 2010b)

In the episode of the British speculative science fiction anthology series, *Black Mirror* entitled Nosedive, the characters live in a world where people rate every interaction they have each other on a scale of one to five for every social interaction they have. An aggregate score that accumulates in real time is shown to these people in real time by a dashboard interface on their smartphones. This score greatly impacts the characters' lives and controls their socioeconomic status. The scores dictate what neighbourhoods people can live in, what types of jobs they can apply for and who they can date. Like many of the themes covered in *Black Mirror*, the episode deals with the anxieties of a society obsessed with subjecting themselves to constant tracking, measurement and being rated by others. Nosedive paints a dystopian portrait of what self-tracking at the level of entire communities and populations could develop into. It also portrays conditions that are already beginning to emerge.

Nosedive's revolves around protagonist, Lacie who works hard at maintaining her 4.2 out of 5 rating though she really wants to be above 4.5. When Lacie is asked to be the maid of honour at a friend's wedding – someone who is highly rated – she sees it as an opportunity to increase her score which would help her move into a nicer apartment in a better neighbourhood. Unfortunately for Lacie, despite her diligence and best attempts to not put a foot wrong, a series of unfortunate circumstances result in her score nosediving. Her brother gives her a one star rating after they have an argument, then, a taxi driver gives her a low score because she's kept him waiting for too long. When her flight to the wedding is cancelled she is unable to rebook because by this time her score is too low to qualify for any customer benefits with the airline. As Lacie's life descends into chaos, her feelings of alienation and exclusion are palpable and we can really get a sense of how high the stakes are when decisions are made over people's lives on the basis of data over which they have little control. Of course Lacie is in control of her actions but she has little agency and the decisions she makes at any given

The Politics and Possibilities of Self-Tracking Technology, 87–101

Published under exclusive license by Emerald Publishing Limited
doi:10.1108/978-1-80043-338-020211004

moment are made against an ever-present stream of data that is accumulating in real time. The measurement of her social interactions guides her behaviour.

This brief example from popular culture makes reference to a world where data-garnering technologies are embedded deep into the fabric of everyday life. And although it is, to an extent, an exaggerated portrayal, the idiosyncratic nature of data-driven sociometric technology and the emotional effects of their usage are real. This chapter examines sociometric technology with the society *Nosedive* depicts in mind. Its focus is on one of the contexts where sociometric technology being rapidly adopted – the workplace. I argue that workplaces are particularly susceptible to the widespread assimilation of sociometric self-tracking because it is a context in which the technology can be presented as unproblematic extensions of other methods of behaviour monitoring and social engineering which are already widely adopted and so common that they are rarely questioned.

Audit Cultures

Self-tracking in the workplace has deep ties to the pen and paper based audit cultures that, since the 1980s had been used by organisations to foster procedural accountability and interpersonal transparency. These qualities were seen as a means of actively cultivating and training personnel into the embodying the traits that an organisation might see as essential to an effective work culture (Cooke & Rousseau, 1988; Deal & Kennedy, 1982; Fombrun, 1983) and organisational learning (Cyert & Kumar, 1996; Schön & Argyris, 1996). These efforts gave rise to an ideology of human resource 'management' that saw organisations learning through individual and collective self-reflexive practices of measurement and tracking. By placing the term 'management' in quotation marks here, I am drawing on the definition offered by Raymond William who in *Keywords* highlights a dual function of 'management' that fits the argument of this chapter (1983/2014, p. 191). Williams describes management as a bureaucracy, and in the context of this chapter's discussion, we could say that this includes technocracy as the work occurring when a select group of elites administer processes of human control in ways that aim to support pre-defined norms that serve managerialism. The other function of management that Williams offers refers to is the abstraction of relations that are embedded in processes that automate and bring about the internalisation of external forms of power and control in ways that would seem as if they *are good for and in the best interests those being managed.* Quantifying organisational cultures and the social cohesion within the workplace, according to this ideology of technocratic managerialism, could be leveraged in ways that could be used to construct a brand-supporting image of an organisation's people and culture. This effort incentivises, or in some case mandates, intense introspection on the minutiae of work life and encourages the subject to draw connections between personal traits – what they might see subjectively as strengths, weaknesses, successes or failures – and larger organisationally set value systems. The assumption here is that healthy organisational cultures are ones that are viewed to be those that invest in cultivating this type of reflexivity and learning where individualism and diversity is valued but these qualities are also something that can be measured, aggregated, scaled and shared with others (see Kotter, 2008). The organisational sense-making that occurs here is one expressed through an internal dialogue that goes something like: *what's being asked of me? what do my behaviours, actions and attitudes convey to my manager or my peers about me?* and *how might my workplace's expectations of me map onto my personal views of what considered appropriate workplace behaviour more generally?* (see Harris, 1994). In trying to answer these questions, an employee might rely on a number of documented resources. Being measured and monitored is viewed as an expected, beneficial, and unchallenged requirement of their terms of employment.

Audit cultures evoke an aspirational language and make it part of everyday corporate discourse. Yet, its use also carries with it a degree of anxiety and expectation that there are likely to be small resistances (Kipnis, 2008; Strathern, 2000, pp. 1–2) and workarounds that unfold in parallel to documented accounts of events. The assumed advantage of auditing is that it opens up channels for productive discourse between the auditor and audited. But that complete honesty and transparency in this discourse is unlikely given the power asymmetries of the relationships between these parties, and in the audit instrument itself. Despite the appeal of applying self-tracking technology to develop better, more precise and less game-able auditing tools, those who are subjected to these systems trade off convenience and ease of use against risks of disclosure, re-purposing and lack of meaningful ownership of their own context specific performance data as an asset which may appreciate in value over time or under certain circumstances. As Andrew Murphie argues

> …[the] documenting and recording of life is now possible with all the extraordinary detail and portability made available by [wearable and cartable technology and] archives, data circulation, and aggregation and the new forms of expression these make available with which to think and act. (Murphie, 2014)

The forms of expression afforded to that are considered to be positive qualitative traits such as mentoring, caregiving, common courtesy, inclusiveness, cultural competency, respect for others and conviviality are reduced to simplistic, often meaningless indicators when they become auditable. This is not surprising given that attitudes to work are shaped by a chain of reasoning that assumes (Adler, 1993, pp. 97–98) the following: when tasks are routine and repetitive the efficient and accurate performance of these tasks requires standardisation in terms of time bodily ability and neurodiversity; high levels of standardisation strip tasks of their intrinsic enjoyment and demotivate the person doing them; engaging in demotivating tasks leads to disaffected and dysfunctional behavior including absenteeism, high attention to work quality, high levels of organisational turnover and obfuscation, interference, deliberate or organised non-compliance and sabotage; counterproductive behavior by workers is best addressed by more rigid and authoritarian management, more hierarchy and higher levels of reporting.

Technologists and social scientists have long engaged in collaborative research dealing with questions of technical intervention into the management of group effectiveness and workplace socialisation, especially of newcomers (Coovert & Thompson, 2013; Flanagin & Waldeck, 2004; McLoughlin & Clark, 1988), employer assessment of worker performance (Reio & Sutton, 2006; Stephenson, McDonough, Murphy, Nugent, & Mair, 2017) and organisational teaching and learning (Schmidt & Kunzmann, 2006). Traditionally, collaboration between social scientists and industry in this area has involved paper-based survey-like instruments or human observation of situated group interaction, but the data obtained by these methods are subjective and open to the introduction of managerial biases when interpreted and reported on. Further, they cannot be analysed in real-time. Similarly, ethnographic methods to study the social dynamics of workplaces (Brannan, Pearson, & Worthington, 2007; Ruhleder & Jordan, 1997; Zickar & Carter, 2010) often produce compromised data because when interviewed in workplace settings informants are likely to be apprehensive about discussing issues of organisational performance or personal non-compliance with candor (Crabtree, Rodden & Tolmie, 2009; Hodson, 1998).

The tools that are used to study workplace sociality, such as surveys, human observation, aptitude testing, psychometry and controlled experiments, however, have a number of problems associated with them. When they are offered as opt-in initiatives, the motivation to complete paper-based surveys or to participate in workshops, focus groups or interviews is likely to only be from those who actively want to contribute to initiatives designed to measure and improve organisational culture. This coupled with other contextual and individual factors introduces a bias into the data that can be collected even if it is anonymised given that the time and labour commitment one gives to such things would assume a certain degree of motivation and willingness to contribute. Even when survey, interview and observational data are supplemented with other forms of data, observer bias is likely to frame the results in a certain way that cannot exactly be assumed to be neutral or objective. As Andrew Murphie notes:

> Any reorganisation of work and everyday life outside of audit will require new cultures and an inventive technics, a technics that enables singular individuations of work experiences and of occasions of living. Finally, it will require a new kind of transversality an establishing of relations that are not restricted to those of auditor/auditee or the measured circuits of financial capital or social control. Of course, it might also require allowing people to get some sleep – their own sleep. (Murphie, 2014)

Auditing, leads us inevitably to worker alienation and employee belligerence. This brings with it the challenge of how one might design workplace performance data capture systems to be more accurate and less intrusive into regimes of technology-enabled performance management. An example is discussed in the next section.

Sociometric ID Badges

Contemporary workplaces, physical or online, are organised to encourage collaboration and inter-functional coordination across different centres of organisational knowledge and value production. Surplus value for organisations is produced in the accumulation of efficiencies in speed or time on the one hand, and on the other, through the interlacing of these efficiencies with an awareness of the actions that have produced these efficiencies and how they might be cultivated, sustained and replicated. Managing tasks, requests deadlines, interruptions, emails, meetings, phone calls and so on in addition to one's own ambition to succeed or merely cope becomes linked to the notion of employee wellbeing. A common challenge for organisations is fostering group collaboration. Group productivity suffers when individuals are thought to be either making less of an effort to contribute or when they compromise productivity by over-participating and monopolising group meetings or when they actively engage in incomplete information disclosure or exchange. Technologists tend to come at these challenges by exploring various ways to track group dynamics in real-time in order to correct or alter workplace behaviour. Self-tracking technology offers the analysis of social cohesion in the workplace such as open plan offices, retail showrooms, laboratories or classrooms a range of benefits. Sensor technology can in some ways be more accurate than self-reporting, less intrusive and quite powerful in terms of the range of variables that they can measure but are they adequate in their capacity to capture the nuances or organisational behaviour?

Technologies for measuring human behaviour in the workplace are usually designed to be introduced by an employer and take the form of unassuming familiar forms such as small discrete sensor embedded technologies encased in a housing resembling an ID card that can be worn on a lanyard which is a familiar sight in many workplaces. Among the first examples were attempts by Olivetti Research in the early 1990s to redesign the inanimate name badges commonly used in organisations to identify themselves and/or to gain access to certain locations (Want et al., 1992). Olivetti's *Active Badge* contained a small simple microprocessor that could broadcast the identity of the wearer and its primary uses were to track the location of wearers in office spaces and to trigger automatic doors. Later, *Thinking Tags* and the *nTAG* systems were designed to show via an LED display how much people at a conference had in common based on a series of standardised questions which users answered prior to an event and to allow them to network and share ideas. By the late 1990s Phillips Research Labs were designing ID badges capable of detecting pre-ambulatory activities using an embedded digital accelerometer. This device, named the *iBadge* was designed to be worn by children to capture their interactions with teachers in classroom settings. The *UbER Badge* developed at MIT Media Lab was an RFID, IR and LED system that was capable of presenting graphics and small amounts of scrolling text in order to facilitate interpersonal interaction in large groups.

By 2009 David Lazer, Alex Pentland and a number of other 'endorsing colleagues' were writing about the potential of computational social sciences to 'increase knowledge of individuals, groups and societies using data with an unprecedented breadth, depth and scale' in *Science* (Lazer et al., 2009, as cited in Pentland, 2014, p. 217). They write:

> To date, research on human interactions has relied mainly on one-time, self-reported data on relationships. New technologies, such as video surveillance, e-mail, and 'smart' name badges, offer a moment-by-moment picture of interactions over extended periods of time, providing information about both the structure and content of relationships. For example, group interactions could be examined through e-mail data, and questions about the temporal dynamics of human communications could be addressed: Do work groups reach a stasis with little change, or do they dramatically change over time? What interaction patterns predict highly productive groups and individuals? Can the diversity of news and content we receive predict our power or performance? Face-to-face group interactions could be assessed over time with sociometers. (Lazer et al., 2009)

The *Sociometric Badge* was one of the devices being referred to on the passage quoted above. It is an electronic sensing device that collects and analyses data on human behaviour and social interaction. Its primary application was to study organisational networks and the flow of ideas within them. Although the existing versions of this device do not record conversational content, they analyse social signals such as enthusiasm, interest, persuasiveness, nervous energy and turn-taking from a combination of speech and biometric data. Using a 3-axis accelerometer similar to those found in most smartphones and wearable activity tracking devices, the badge measures body movement including walking, sitting and other gestures associated with social interaction including mimicry and rhythmic motion. The proximity between wearers is determined by using RF or Bluetooth and face-to-face interaction is detected by an infra-red sensor calibrated to detect facial alignment between people and detect postural position of users relative to one another. Penland describes the badges and their accompanying analytics platform as a 'longitudinal living laboratory, or social observatory' coupled with a support system infrastructure that enables the sensing and collecting of data, data processing, and a set of tools for feedback and communication with the subject population (2014, p. 218).

Ben Waber, one of the MIT-based researchers who developed the technology with Pentland, describes the badge in his book *People Analytics* (2013) as:

> …the natural evolution of the company ID badge. No longer just a tool to open doors, this new kind of ID badge enables you to understand yourself and your company at large through data-driven reports and feedback […] The sensor technology has amazing potential. From gathering five minutes of data with the badge, you could figure out not only whether you'll win a

> negotiation, but how well you will do. With this badge deployed across millions of individuals at different companies all over the world not for minutes but for decades imagine what we could learn about to help people collaborate more effectively and create better organisations. (Waber, 2013, p. 16)

The *Sociometric Badge* measures and correlates a number of variables including speaking time, average speech segment length, variation in speech energy and variation in movement (Kim et al., 2008, pp. 461–462). It is not hard to understand the appeal for this technology at the enterprise level primarily but also by smaller groups within an organisation as a proxy for constructive feedback on performance and for the purposes of reward and recognition. The data from the badges is likely to be linked with data from CCTV cameras, key-logging software, automated randomised screenshots of employee computers and data from email correspondence. As a reaction to this, those using the badges may also demand, or seek out, methods to increase the degree of transparency around their own performance and that of their peers. The *Sociometric Badge* could be used to establish standards which if not met, or exceeded, to trigger performance management measures being imposed on the user which could be effective in motivating positive behavioural change though it is likely that the standards themselves will serve organisations more than employees. But long-term use of the *Sociometric Badge* is also likely to be accompanied by feelings of overwhelm when morale suffers and users grow tired of the speed, complexity and intensity of information flows about their performance that govern the social norms of their workplace. When management wants to convince workers there is a hidden but measurable dimension to their performance that requires the use of technology for it to be measured and analysed, they are harnessing, and re-directing the energies an individual invests in their own self-betterment and professional development.

When Pentland and Waber formed a Boston based consultancy founded called, Sociometric Solutions to market the *Sociometric Badge*, they offer it to potential clients as a solution that can do away with the problems associated with:

> …surveys and human observers [which] provide biased, limited views into communication behaviours, which is of little practical usefulness for organisations. We focus on measuring real-world social behaviour using Sociometric Badges. These wearable electronic sensors are capable, among other things of: Capturing face-to-face interactions; extracting social signals from speech and body movement; measuring proximity and location of users. (sociometricsolutions.com)

Although the *Sociometric Badge* uses voice recognition technology that measures tone, frequency, inflection, and cadence it is not set up to capture data about the content of conversations. Data on tone, frequency and cadence is used in combination with behavioural and physiological data to feed models for speech pattern analysis that can measure conversational styles and map that against predetermined criteria to score users on their communication skills. The *Sociometric Badge* extracts data on energy expenditure, levels of engagement at the personal and aggregated level. When this data is linked and cross referenced, the patterns that emerge are used to make inferences about social cohesion within groups, motivation levels, stress or even illness. Bringing together a range of data on all of these factors also allows for calculations to be made about individual levels of introversion, extroversion and empathy (Pentland, 2014, p. 221). The *Sociometric Badge*, in conjunction with a dashboard that reports analytics including activity level and mirroring, along with feedback reports that show how an individual's behavior compares to those around to aid managerial and self-identification of areas to be addressed in individual or group based personal or professional development programs. Combining the data that it captures with non-sensor data sources such as a subject's electronic communications and other objective productivity metrics in an integrated visualisation tool that displays data from the sociometric badges insights into behaviours such as turn-taking, speaking time, speech transition and dominance can be provided to managers and team members.

Applications for wearable sociometric technology include the analysis of employee-customer interaction in retail and consultative selling, studying structural and cultural difference in organisational mergers and acquisitions, training and personal development, organising workplace infrastructure and layout and understanding team dynamics in collaborative work. The *Sociometric Badge* has been used for research and proof of concept studies to study social dynamics in creative agencies, hospitals, banks, warehousing and call centres (Pentland, 2014, p. 220). The vision for this technology, according to its proponents is that it affords the analysis of multiple modalities of an individual's working environment as a combination of directly sampled and self-reported data in order to investigate 'natural and externally imposed social mechanisms relating to behaviour and decision making, and designing and evaluating new mechanisms and tools to help people make better decisions' (Pentland, 2014, p. 223). But there are other applications for *Sociometric Badges* and devices like them in situations where interpersonal activity might require official supervision or mediation. This raises ethical concerns over the extent to which a device that is recording social interaction even when it is being used by one participant in a group should be disclosed to others in the same way that audio, or video recording is. Peer pressure and obligation, informed consent, individual control, and anonymity in the form of aggregation when access to the data is granted to third parties would need to form the basis of stringent guidelines for handling data produced through digital sociometry, but even these protective measures still frame self-tracking in terms of the mitigation of organisational risk as opposed to protecting users from the psychological effects of having social interaction measured and tracked.

The question of how such data is to be covered in employment contracts in the sense of who actually owns the data is likely to become increasingly relevant in terms of the rights and responsibilities associated with accessing and repurposing the data. For instance, could an outgoing employee use the data in demonstrating capacity in addressing selection criteria when applying for positions outside the organisation who administered the sociometric analysis? How might such data be used when giving or seeking out employment references? What internal circumstances might trigger an embargo on the use of an employee's data? For example, on the *Sociometric Badge*, Kitchin notes:

> ...the analysis is riddled with assumptions of social determinism, as exemplified by Pentland (2014) 'the sort of person you are is largely determined by your social context, so if I can see some of your behaviours, I can infer the rest, just by comparing you to the people in your crowd'. In contrast, human societies, it is argued, are too complex, contingent and messy to be reduced to formulae and laws, with quantitative models providing little insight into phenomena such as wars, genocide, domestic violence and racism, and only circumscribed insight into other human systems such as the economy, inadequately accounting for the role of politics, ideology, social structures, and culture (Harvey, 1972). People do not act in rational, predetermined ways, but rather live lives full of contradictions, paradoxes, and unpredictable occurrences. (Kitchin, 2014, p. 8)

A problematic assumption in the analysis of badge data is that people stay on topic when they are speaking in meetings, unless the badge is configured to record the actual conversations or people alter their behaviour in order to be less of their natural social selves in meetings on account of the device. The software used to analyse data from the badges affords a greater capacity to perform analyses on employees outside the traditional method of objective or target setting and measuring progress towards that benchmark given the extent to which they can allow a given indicator or measurement to metonymically stand in as a marker of some or another given characteristic, for example voice and speech characteristics as markers of dominance or aggression, eye tracking as a measure of concentration or keystrokes for productivity. The integration of biometrics and facial recognition software into self-tracking systems may see workers measured on physiological variables such as smiling (Gawne, 2012) or truth telling which brings about the important question of where might practices of resistance to self-tracking in the workplace be situated given that refusal can to some degree be viewed as choosing dissonance resulting in workers making themselves thrive organisationally only to become anxious about their performance. If self-tracking is to be thought of as an emancipatory practice then it achieves this only though certain permitted and pre-programmed forms of mediation and measure which reduce self-knowledge and its accompanying potentials of empowerment and resistance to a network of relations governed by a logic which alters workplace self-reflexivity, and forces a type of reflection that is far removed from that which would ordinarily occur when an employee is being asked to answer questions about their lived experiences.

Given the complex nature of these implications for the work/life boundary, workplace sociometry is likely to evolve into an occupational health issue as well as a political one. In response, those participating in self-tracking at work may also demand, or seek out, methods to increase the degree of transparency around the analysis of their own performance data and that of their peers as a form of inverse surveillance. The panopticon metaphor is often evoked as a point of departure in discussions of the surveillant tendencies within the technocratic management of everyday life. As an architectural model, and by extension a material technological presence in people's lives, as a model the Bentham/Foucault panopticon demonstrates how human information flows can be structured with asymmetries that result in the embodiment of forms of self-tracking where there is an ambivalence towards the physical presence of someone or something mediating that information. In their study of Closed-Circuit Television (CCTV) funded by the Danish Council for Independent Research which sought to examine the everyday contexts of participatory surveillance (Albrechtslund & Lauritsen, 2013), they argue that surveillance and its effects are a matter of aligning heterogeneous actors and how their participation is negotiated is what shapes the ways in which intentions, and power, are translated. They argue, is to analyse any given situation where surveillance occurs which includes the way participation is established. In the context of sociometry, this is even more difficult. Albrechtslund and Lauritsen state:

> …an important task for surveillance scholars is not only to identify participatory surveillance as a specific iteration of surveillance. Even though ANT [Actor Network Theory] is to some extent inspired by Foucault's work, this does not entail an explicit focus on power. Rather, the project is to analyse participation in any given situation of surveillance, and this includes a careful attention to how participation is established, maintained and negotiated. (Albrechtslund & Lauritsen, 2013, p. 311)

As one commenter on Gary Wolf's (2010b) *New York Times* article *The Data Driven Life*, suggests:

> Just as with any other system of data/records (accounting springs to mind) human behaviour makes us prone to (1) creating or finding loopholes and workarounds or concentrating on one aspect while ignoring others (2) concocting mis-interpretations and mis-applications of data results (3) the development of uneven or inconsistent standards (in personal record-keeping, perhaps, 'must have' results sets developed and marketed along with the sensor package). Meticulous personal/social record keeping may offer some insights into behaviour, and may develop new schools of thought on many issues, but it will create its own entire set of dis-functions, misconceptions, misuses

> and mistakes. On another note - who's going to provide to the masses the training in the analytical skills necessary to understand and interpret their data? Gives an interesting twitch to 'you do the numbers'.

In workplaces a degree of surveillance and monitoring is expected but the extent to which a sense of intrusion might be felt with the extension of this into personal spaces and times is subjective. While the *aspect* of surveillance is not in any way obscured it is the *extent* of that surveillance that can be strategically obscured by those conducting the surveillance. This means that the extent to which workers might adjust and regulate their behaviour to work with, or around sociometric technologies is based on an assumption that the exact nature of tracking has been adequately disclosed. Take for example the claim that the *Sociometric Badge* does not record the content of conversations. It could be programmed to do so and a number of ways of structuring conversational data might be deployed. Even in the instance that aggregated conversational data is used the nuances of individuals' linguistic patterns could, in theory, be used to re-identify individuals from de-identified datasets.

Two main perspectives can be linked to two complementary expressions of data-driven surveillance in the workplace. Firstly, the individual use of various techniques of self-surveillance and tracking. Second, the simultaneous trend to optimise urban infrastructures through smart information technologies which link agency and the impact of the outsourcing of meaning as an act of data interpretation on human agency within frameworks of external, automated surveillance. The type of participatory surveillance that the *Sociometric Badge* brings with it is a 'specific iteration of panoptic surveillance' (Albrechtslund & Lauritsen, 2013, p. 311). The nature of social surveillance afforded by the badge is one that connects to Foucault's paradox of a 'conduct of conduct' where the badge is programmed to capture and identify productive or good patterns of individual behaviour and feed them into models designed to scale them up and deploy them outwards as best practice within a given organisational context while at the same time requiring that individual behaviour be modified in the first place in order to account for the presence of the device.

Legal scholar Jake Goldenfein notes that 'the relevance of the panopticon as a metaphor begins to wither when we start thinking about whether contemporary types of visuality (digital and data-driven) are analogous to the central tower concept' (2015). Unlike the tower, the embodiment of badge-surveillance presents a range of strategic options that might be available to the user. For example in their analysis of surveillance and information flow symmetry, Jonah Bossewitch and Aram Sinnreich (2013) which develops a range of 'sociographic heuristic models' for the various ways that people self-present in ubiquitously monitored environments we see an architecture for post-panoptic surveillance through which the design of sociometric self-tracking systems can be critiqued. Looking beyond the surveillance/counter-surveillance conception of power symmetry, a model Bossewitch and Sinnreich's (2013, pp. 230–236) is helpful. It is summarised and developed in Table 4.1.

Table 4.1. Information Flows in Surveillant Data Transactions.

Model	Pattern of Data Transaction
Total transparency	Often touted as a solution to asymmetric configurations of power but significantly reduces individual privacy. Asymmetric forms of power are likely to be reinstated on the basis of differential access to personal data and the resources available to individuals and third parties to process, interpret and act on the data.
Going off grid	Making an effort to disappear as much as possible from some or all regimes of data collection and reduce data exchanges to the most minimal amount possible. Difficult to achieve in any meaningful way given the dependence on mobile phones, internet connections and needing to have a valid email address and/or credit card to access services.
Creating a black hole	A less avoidant mode of data transaction where the subject is part of, and uses, data from within the regimes they are part of while letting out personal data in a highly selective or minimal fashion. This strategy assumes a high degree of technological and informational literacy.
Promiscuous broadcasting	Can be practiced strategically or non-strategically by producing large amounts of information that gives the appearance of total transparency along the lines of having 'nothing to hide' but can also be used to confuse outsiders that might be ill equipped to make sense of the sheer volume of raw data.
Voracious collecting	Collecting data on a range of variables on a single platform or multiple platforms, some for a given purpose and some may be for stockpiling for some future use of evidentiary purpose.
Spreading disinformation	Actively propagating disinformation so that outward flow of accurate information about the subject is reduced. A strategy laden with risk in terms of how accountable one might be made about the data they disclose and the level of truth that third parties might ascribe to inaccurate data that is produced.

On such tendencies, Waber notes:

> This type of environment completely defeats the purpose of the badges and organisational sensing in general. The badges are designed to help make people happier and more productive, things that all companies should strive for. There's just not a good business case for checking what Bob was doing at 2:30 p.m. on Tuesday. Not only is exposing that information a huge violation of privacy, it also wastes the company's time. The things that all companies should care about are: what makes people in this company productive and happy? [and] How can the company change to make people happier and more productive? (Waber, 2013, p. 180)

But Waber downplays the fact that interpersonal factors shape how business information is used or that workplace bullying, and harassment are very real problems and something that is readily experienced by most people at some point in their lives. And although there are benefits of sharing aggregate data across sectors, what is being ignored here is what monitoring means for the people being monitored. Pentland, on the other hand, suggests:

> …this vision of a data-driven society implicitly assumes that the data will not be abused. However the ability to see the details of the market, of political revolutions, and to be able to predict and control them is a case of Promethean fire – it could be used for good or for ill. In short, to achieve the exciting possibilities of a data driven society we require workable guarantees that the data needed for public goods are readily made available while at the same time protecting the citizenry. (Pentland, 2014, p. 17)

When sociometric technology is designed it is probable that not all of the archetypal behaviours listed in the table below would be considered equally (Table 4.2).

The way that technologists see the social as something to be measured is, however, somewhat problematic. In *Reassembling the Social: An Introduction to Actor-Network Theory*, Bruno Latour captures the essence of the problem capturing data that observes daily living and social reality by technologists being too narrow and simplistic. He writes:

> …when social scientists add the adjective 'social' to some phenomenon, they designate a stabilised state of affairs, a bundle of ties that, later may be mobilised to account for some other phenomenon. There is nothing wrong with the use of this word as long as it designates what is already assembled together, without making any superfluous assumption about the nature of what is assembled. Problems arise, however, when the 'social' begins to

Table 4.2. Proximities to Real Selves in Sociometric Data Collection Initiatives.

Strategy	Behaviour
Avoid	Assumes the choice to divulge or contribute information.
Protect	Actively takes steps to secure their data or to understand the purposes of its collection and what their rights are.
Accept	Active engagement and compliance with the data collection regime, likely to trade concerns about liberty and privacy off for incentives, norms or other pragmatic attitudes to the collection of data.
Contribute	As above but with advocacy for the collection regime. Sometimes this may be the result of having a personal connection or vested interest in the type of data being contributed.
Obfuscate	Like protection and avoidance but the individual may deliberately falsify data rather than outright reject or opt out. Wanting to collect an incentive may be a factor here.

> mean a type of material, as if the adjective was roughly comparable to other terms like 'wooden', 'steely', 'biological', 'economical', 'mental', 'organisational' or 'linguistic'. At that point, the meaning of the word breaks down since it now designates two entirely different things: first, a movement during a process of assembling; and second, a specific type of ingredient that is supposed to differ from other materials. (Bruno Latour, 2005, p. 1)

Attempts to track and measure social interactions suffer from this exact problem of designation. In an article titled, *On San Francisco's Quantified Self Delusion*, sociologist, and self-labelled 'documentarian of the absurd', Holly Wood writes:

> …we live in a neoliberal dumpster-fire of an epoch where everyone insists everything must be objectively measured. And once things are measured, they can be objectively ranked. And when things are ranked, then rewards can be objectively distributed. And then when people suffer, assholes can claim everything is justified because we distributed resources based on rankings which in turn were based on measurements (and this is tautological as fuck). (Wood, 2016)

Sociometry does little to represent human social action and in what follows, I want to show how sociometric technologies continue a trajectory where personal data 'express[es] power relations, symbolise social hierarchies, reinforce social inequality, transport social power, objectify inequality' (Latour, 2005, pp. 70–72). The normalisation of monitoring and measuring behaviour in the workplace has played, and will continue to play a role in, the transfer of sociometric techniques across the work/life boundary.

As the Solidarity Federation, an anarcho-socialist section of the International Workers Association wrote:

> We need to forget the stuffy and bureaucratic image of health and safety. The result of the erosion of working conditions and training means we take more risks and work more dangerously – the statistics speak for themselves. We don't have anyone else who we can trust to deal with the issue, but ourselves. This may just seem like fine words, but once you start looking critically at your own workplace, and how it could be changed to suit workers, the view of the self-management process snowballs. It becomes clear that if we could control our workplaces ourselves, and if we were running them for people's needs, then they would look very different.

In this chapter, I have argued that the logic that underpins the *Sociometric Badge* assumes that 'human societies are mostly made of networks of exchanges between individuals' (Pentland, 2014, p. 21) but it cannot be denied that no matter how altruistic the vision for sociometric technologies are, they produce dis-empowering subjectivities and these matters are important considerations as priorities in the design of organisational sociotechnical systems and as a matter of occupational health and safety and of maintaining a boundary between life and work.

Part 3
The Quantised Self: Discipline and Design

Chapter 5

The Quantised Self

> We tolerate the pathologies of quantification – a dry, abstract, mechanical type of knowledge – because the results are so powerful. Numbering things allows tests, comparisons, experiments. Numbers make problems less resonant emotionally but more tractable intellectually. (Wolf, 2010b)

> We acutely feel the alienation of our temporal selves is borne out in the growing debates on the 'lack' of time in our society. However, it is not enough to centre the debates, [...] on the workplace, or the need to bring some equity to the 'life-work' balance. These are important, but they are effects as opposed to causes of our time-starved lives. (Hassan, 2011, p. 197)

> As with every genuine art, this will not be reducible to a few cheap formulas, a few gadgets to help us organise our time, our comfort, or our pleasure more efficiently. Recipes and techniques for increasing happiness and pleasure are part of the baggage of bourgeois wisdom – a shallow wisdom that will never bring satisfaction.

In the last chapter, I examined how the capture and analysis of data about interpersonal interaction brings about social engineering in the workplace. In this chapter I look further into this tendency but with an emphasis on the body, and how *it* is conditioned to align itself to the rhythms and flows of data and devices. As outlined in the book's introduction, the quantised self is a metaphor and logic that extends this to digital mediation. In this chapter, I pick up on this idea and unpack it in detail. I argue that, Henri Lefebvre's writings on Rhythmanalysis can be developed as framework for critiquing the controlling and aligning aspects of the user experience of self-tracking technology. Published posthumously, Rhythmanalysis is not a straightforward concept. It is also not a straight forward exercise to develop it in new directions, especially Lefebvre's positioning of technology and its products – specialised fields of knowledge and vocabularies, its outputs in the form of informational categories and units of measure. It requires bringing together some incongruent conceptual frameworks and ideas and taking some degree of interpretative licence with these as well as Lefebvre's original text. Consider the unmasking of conditions in factories depicted in the film *Modern*

The Politics and Possibilities of Self-Tracking Technology, 105–121

doi:10.1108/978-1-80043-338-020211006

Times (1936), where Charles Chaplin offered a sympathetic and satiric interpretation of rhythmic and mechanical logic entering into everyday life. The film is structured into four segments, each set in a factory, prison, department store and dance hall. In the segment that is set in a factory, Chaplin is seen entering the gates with his fellow workers like a flock of sheep. They commence working. When Chaplin can no longer cope with the 'monotonous regularity of the assembly line' (Tolich, 1992, p. 344) and the increasing acceleration of the cogs and wheels which make up the machines he is meant to working on, he antagonises his fellow workers and managers who chase him through the factory. The scene ends with Chaplin being taken away frantic and berserk in an ambulance. In evading his pursuers, Chaplin uses 'technical control' (Tolich, 1992) to stall them, while being chased, each time he passes the lever that controls turning on or off the assembly line, he activates it and the workers are shown to return to work dutifully, halting their pursuit as if their bodies and behaviour are governed by the running of the conveyer belt. Managers are seen to be in control dictating when production starts and stops. The workers are depicted as having no free will of their own, except Chaplin.

Let us look at a more contemporary example of productivity management before going further. At the QS European Conference in May of 2014, Brian Crain spoke about the *Pomodoro Technique* a start/stop method which he had been using to track his productivity and time he spent on particular work-related tasks. Crain explains:

> I learned that having a continuous metric is enormously motivating since it allows you to continually improve yourself. These small, continuous changes make a huge difference over time. I also learned that building a user-interface is tricky, but very important to make tracking rewarding. This is something I successfully did with the Pomodoro Technique, but have found difficult to replicate with other methods. Finally, tracking commitments has taught me how critical one's mindset is. When I would slip into thinking of commitments as simple tasks, my success with that method derailed completely. So, for that method, I realised how important it is to build a system and user interface that helps maintain the commitment mindset. (Ramirez, 2014)

The technique, which revolves around creating a rhythm between working and relaxing throughout the workday, was developed by Francesco Cirillo (2006) as a means of focusing his concentration during his university studies in the late 1980s. The technique aims to address the anxiety of time pressure and procrastination that results from the stress of working against deadlines. Drawing from Bergson and Eugene Minkowski, Cirillo (2006) describes in a manifesto written about it:

> If we try to measure ourselves against the passage of time, we feel inadequate, oppressed, enslaved, defeated more and more with every second that goes by. We lose our *elan vital* [vital impetus]

> our vital contact which enables us to accomplish things. In a moment of weakness, the purpose of the activity is no longer even clear. The succession of events, instead, seems to be the less anxiety-ridden aspect of time. At times it may even represent the regular succession of activity, a calm inducing rhythm. (p. 3)

The basic method is to plan out a set of activities for the day and collecting 'raw data on the effort expended and other metrics of interest', recording at the end of each day an archive of these daily observations and reflecting or processing this data into insights on the day's work by visualising the 'information in a format that facilitates understanding and clarified paths to improvement' (Cirillo, 2006, p. 5). The technique is executed by using a timer to time twenty-five-minute increments (Pomodoros) which are designated 'indivisible' slices of time that if interrupted are void. After an uninterrupted 25 minutes a short three to five-minute break is taken to disconnect from the work-task, reflect and breathe, stretch, walk or daydream. When the timer sounds to announce that 25 minutes has past, Crillo asserts that work must absolutely stop, there is to be no working for 'just a few more minutes' (2006, pp. 6–7). Crillo advises:

> During [a] quick break, it's not a good idea to engage in activities that call for any significant mental effort. For example, don't start talking about work related issues with a colleague; don't write important emails or make imperative phone calls, etc. Doing these kinds of things would block the constructive mental integration that you need in order to feel alert and ready for the next Pomodoro. You should include these activities in your Activity Inventory and earmark specific Pomodoros to do them. Clearly during this break, you shouldn't continue to think about the work you've done during the last Pomodoros. Once the break is over, set the Pomodoro [timer] to 25 minutes and continue the activity at hand until the next time it rings. (Cirillo, 2006, p. 7)

Rhythmanalysis emphasises observation, apprehension and representation of rhythms that exist in the world and are then subsequently given form through the mediating capacities of instruments and technologies. The rhythm of learning associated with Pomodoro technique for instance can be understood in terms of a 'triad' (2004, p. 41) that Lefebvre references to discuss his rhythm-analytic model of training: (1) an internal activity that can be controlled under some form of direction that is 'punctuated by pauses for repose'; (2) completely stops for rest or sleep; and (3) diversions and distractions that accommodate the recommencement of the cycle of training, a reward of some sort (2004, p. 41). Before Lefebvre, Gaston Bachelard developed theories of the self that borrowed from physics privileging the undulatory nature of how time is experienced. According to Bachelard, time was not the objective experience of incremental measured units passing as one observes by checking the time on a watch or the date on a calendar. Rather, the subjective experience of time arises from a capacity to relate and cognitively assemble successive and discontinuous instances which amount to an individual's life experience. The notion of continuity throughout one's life is a series of artificial and tensions or disturbances – such as successes and failures, remembering and forgetting – are needed for the conceptualisation of self. (Bachelard, 1950, p. 142). Rhythmanalysis, for Bachelard aims to 'find patterns of [this] duality' (1950, p. 141). A duality that is apprehended as rational and intuitive and when embodied has some type of healing effect on the individual. Bachelard saw this as a self-motivated reflective practice where an individual practiced self-analysis through the use of media such as literary works that help grasp one's own subjectivities by interpellation (Sauvanet, 2000, p. 107).

For Lefebvre, everyday life was a source of alienation. His development of rhythmanalysis outlines an approach for understanding these alienating forces through the apprehension of biological, physical and social rhythms by the human sensory apparatus. This cognitive and somatic apprehension of rhythms are, for Lefebvre, a gateway into understanding the political and economic forces that govern and shape human subjectivities of the self, and the world, under industrial capitalism dialectically. Rhtyhmanalysis can be broken down as such. At its core there is the idea that the study of rhythms constitutes a type of science where the individual subject is the point of origin in an epistemology founded on the analysis of everyday patterns and routines. Rhythmanalysis rests on a process of ontology building, what Lefebvre calls 'thingification'. This, in his conceptual framework is a dialectical configuration of oppositions that bridge the divide between the quantitative and the qualitative. The idea that at the heart of the rhythm analytical project is that there is a central protagonist, 'the agent', or 'the rhythm-analyst' and that this individual has certain methodological obligations to the project. Foregrounded attention on one type of rhythm results in a training and disciplining of the mind and body (*dressage*). For Lefebvre, rhythms exist at the intersection of place, time and the expenditure of energy and in order to observe rhythms outside the body the rhythmanalyst must use the body and its sensory organs as a way to apprehend rhythms. A rhythm within this particular framework is understood as the combination of the rhythmanalyst and the rhythm being studied as a mutually constitutive entity. Within this framework, Rhythmanalysis attempts to bridge the divide between qualitative and quantitative measures of everyday ebbs and flows so that it can cross various disciplinary domains to illustrate how the embodiment of various rhythm as become second nature linking the subject to a logic of relations between time, space, and body. Being aware of these relations is not always easy because the function of these logics is to make certain rhythms and their states of awareness to become taken for granted (Edensor, 2011).

Rhythmanalysis also bridges embodied social relations, individual self-awareness and the mutual shaping of social, spatial, temporal and biophysical subjectivities as a series of patterns which occur in or out of phase with one another which can be demonstrated and given medial form in time-lapse photography of bodies moving through space for example (see Lyon, 2018; Simpson, 2012; Tiwari, 2010). In rhythmanalysis, the body is considered as central and metronomic relative to the environment in which is it is situated. It is subject to (and subjected to) normal and disruptive rhythms of acting on its physiology, psyche, cognition and proprioception. When the body is represented through the specialised epistemologies of healthcare, labour, gender and so on this results in a body that has been progressively broken down into component systems and then reunified only to be considered whole again in terms of these forms of knowledge where states of wellbeing or productivity are framed by the work and training that needs to occur for these to be achieved. This is a 'skilfully utilised and technicised' (Lefebvre, 2013, p. 47) form of mediation that is has accurate semblances to the thing it represents. But the image that results when movements and behavioural patterns are measured 'takes care of ideology: [in that] it contains it and masks it' (Lefebvre, 2013, p. 47).

Rhythmanalysis is particularly suited to critiquing self-tracking because of the way it sees an 'affective field [of situated and embodied instances of self-tracking in everyday life] so as to render it more predictably [understandable], transgressing the mind-body dualism from the side of the mind and eliminating possibilities for experience except as efficient, rational 'managed subjects' (Moore & Robinson, 2016). It is the emphasis placed on the representation of the quotidian aspects of life, the modulation of qualitative to quantitative forms of representing the same thing, and the application of subject-centric forms of sense-making formultiple, often competing rhythms that makes rhythmanalysis a unique lens for self-tracking technology too. For instance, Lefebvre notes 'technologies kill immediacy (unless speed pass[es] for a return to the immediate but that isn't saying much). The impact of technological conquests does not make the everyday more alive, it nourishes ideology (2004, p. 53).

In Rhythmanalysis, rhythm brings with it four formations each to which Lefebvre attaches a prefix: poly-rhythmia, eu-rhythmia, arrhythmia and iso-rhythmia (2004, pp. 16 and 68). The different types of rhythms and how they be extended to self-tracking can be described as follows. *Poly-rhythmia*, in everyday life, one is confronted with bundles of different rhythms sometimes some of these are in accordance or in harmony and other times they may be out of phase. The ability to grasp these rhythms is event and situation driven. This could state could be understood as an unreflective or superficial, 'first' (Lefebvre, 2013, p. 16) apprehension of the body emplaced in some situational or environmental context. *Eu-rhythmia*, rhythms that are united in a state of health or homeostasis. The discordance of these rhythms signals the arrival of a sub-optimal state. *Arrhythmia*, a pathological or non-normative state that is at the same time a cause, symptom and an effect of itself. *Iso-rhythmia*, the interaction between the rhythmic states of different bodies. One thing that can be gathered from the harmonisation of bodies which are striving to become eudaemonic is a fetish for technology that makes measurement more accurate is a privileging of eu-rhythmic states (Lefebvre, 2013, p. 20). Under eu-rhythmic states, organs and systems are conceived as being in state of a meta-stable equilibrium which when disturbed results in a sub-optimal state of wellbeing.

For Lefebvre (2013, pp. 17–18), physiological rhythms are *kept in secret*, as are the psychological rhythms of things like memories, recollections, conversations and social encounters because of their relation to linear time and memory. The outwardly observable social rhythms, for Lefebvre, are those governed by structured allocations of time and space for instance by a calendar filled with appointments or the patterns of a daily commute or those that are intentionally or unintentionally *expressed* such as hunger or tiredness. Learning processes, for Lefebvre are a form of false rhythm involving short, medium and long-term calculations and 'pseudo-dissimulations' of the body (2004, p. 18). Subjects are dominated or can exert dominance over others by long lasting artificial rhythms which have far reaching effects. So, for Lefebvre, this is the ability to:

> [...] study and compare cases: the rhythms of the body living or not (respirations, pulses, circulations, assimilations â€'durations and phases of these durations etc). This remains close to practice: in confronting the results, the scientific and/or philosophical spirit should arrive at general conclusions. Not without risks: the leap from the particular to the general is not without danger of errors of illusions, in word, of ideology. The other procedure consists in starting with concepts, defining categories. Instead of going from the concrete to the abstract, one starts with full consciousness of the abstract in order to arrive at the concrete. (Lefebvre, 2013, p. 5)

These two procedures of going from the concrete to the abstract and from the abstract to the concrete capture the tension between different rhythms in a way that situates the body not as an objective source of rhythms or truths that can be read off with instruments and interpreted through specialised epistemological frameworks. Rather, they present as conflicted, dialectical, negotiated and ideologically influenced practices where two different modes of knowledge production are brought together and set in motion in a way where they are trying to complete one another *around* a the subject. As Lefebvre writes:

> the second method does not exclude the first: they complete one another. *Specialists*, doctors, physiologists, geologists and historians tend to follow the first, without always arriving at ideas and conclusions that are valid for all rhythms. Here, we follow second, more philosophical method, with its attendant risks: speculation in the place of analysis, the arbitrarily *subjective* in the place of facts [...] Rhythm appears as regulated time, governed by rational laws, but in contact with what is least rational in human being: the lived, the carnal, the body. Rational, numerical, quantitative and qualitative rhythms superimpose themselves on the multiple natural rhythms of the body (respiration, the heart, hunger, thirst etc) though not without changing them. (Lefebvre, 2013, p. 4)

This 'panoply' (Lefebvre, 2013, p. 9) of rhythms: repetition and difference; mechanical and organic; discovery and creation (of self, environment); the cyclical and the linear; continuous and discontinuous, and of course the quantitative and qualitative are measured and brought into representation against one another making the apprehension and representation of rhythms a 'measuring-measure' which is, for Lefebvre the 'depths of the dialectic' (Lefebvre, 2013, pp. 8–9). And, it is this that can considered to be a pathway *out* of the epistemic-political symmetries of knowledge transactions between individuals and so-called experts or specialists. The commodification of one's time, space and energy leaves three simultaneous states of human experience. Firstly, feeling good or not so good about doing things which are being measured. Secondly, the level of coherence and non-coherence between things that are measured. Thirdly, a cognition of the extent to which abstracted, measured forms of real things can be acted on. The type of heightened awareness that results out of the accumulation of these states alters the experience of time as something that has a bi-directionally and linear relationship to tasks performed by the body in both time and space 'if yes, does this operation generate errors, or does it, on the contrary stimulate knowledge at the same time as practice?' (Lefebvre, 2013, p. 10). On the linkage of rhythm, measure and things; times, spaces, and bodies, Lefebvre writes:

> We know that a rhythm is slow or lively only in relation to other rhythms (often our own: those of our walking, our breathing, our heart). This is the case even though each rhythm has its own specific measure: speed, frequency, consistency. Spontaneously, each of us has our preferences, references, frequencies; each must appreciate rhythms by referring them to oneself, one's heart or breathing, but also to one's hours of work, of rest, of walking and of sleep. The preferences measure themselves; the measure (motion and practice) passes through a frequency. Precise techniques enable us to measure frequencies. (Lefebvre, 2013, p. 10)

The idea that 'we *contain* ourselves by concealing the diversity of our rhythms: to ourselves, body and flesh, *we* are almost objects' (Lefebvre, 2013, p. 10) leads to the open question: can contradictions be articulated in technical, quantitative logic without contradiction?

At the heart of the rhythm-analytical project there is an 'agent' (Lefebvre, 2013, p. 18). The analyst-figure who is the subject and also *part* of the object of analysis. The notion of a future self that can be predicted and intervened with or changed through the apprehension of rhythms is one likened to painting 'a portrait of someone who does not yet exist' (Lefebvre, 2013, p. 19) consistently engaging in simultaneous apprehension and anticipation to render form upon this version of the self in formation.

> The rhythmanalyst will have some points in common with the psychoanalyst, though he differentiates himself from the latter; the differences go further than the analogies. He will be attentive, but only to the words or pieces of information, he will listen to the world, and above all to what are disdainfully called noises, which

> are said without meaning, and to murmurs, full of meaning and finally he will listen to the silences. (Lefebvre, 2013, p. 19)

This is to 'acquire knowledge' in discourse. Lefebvre places the word knowledge in quotation marks in recognition of the difficulty in knowing if this type of knowledge can go as far as to be thought of as a science and consequently whether it avoids ideologies, interpretations and speculative constructions which he sees as inevitable when it comes to information technology (2004, p. 22). If for Lefebvre, the present is represented in images, 'the simulator of the present [is] the image (2004, p. 22)' and there needs to be confidence in this image and this seeking of objectivity or at least scrutiny in the image has become a social, aesthetic but not moral obligation that 'gives rise to abuse' (2004, p. 23). As Lefebvre writes:

> Every word, every gesture constitutes an act, and acts must be understood according to their purpose [...] and not merely in terms of the person speaking or acting, as though he could somehow express or externalise his reality and sincerity. More exactly, words and gestures, express an action and not some readymade internal reality. (Lefebvre, 2013, p. 155)

Technologies that mediate this relation will either project a structural logic of units, measures and categories that makes the body a rational unit over which calculative operations can be performed or they aestheticise the body in ways that ascribe social *and moral* obligations (Lefebvre, 2013, p. 23) as part of the art of living' or both at the same time. The rhythmanalyst is meant to be a rational actor, concerned with recovering what can be sensed and deploying (Lefebvre, 2013, p. 25) it almost therapeutically, an empiricist but without a critical declared intervention in mind. The rhythmanalyst:

> ...could, in the long term, attempt something analogous: works [*oeuvres*] might return to and intervene in the everyday. Without claiming to *change life*, but fully reinstating the sensible in consciousness and in *thought*, he would accomplish a tiny part of the *revolutionary* transformation – without any declared political position. (Lefebvre, 2013, p. 26)

Those participating in self-tracking embody a similar ideology. But at the level of praxis there is tacit positivism, declared or undeclared to the position that pursuing forms of strategic agency within the conditions of living laid out by forces of technocracy and neoliberalism 'the revolutionary transformation is veiled and individualistic, the sensible is not reinstated but rather, it is embedded by external forces already set up for its discovery and deployment. But where the introduction of technology into the rhythm-analytical projects — and I would classify the Quantified Self movement as one — is that they become an issue in terms of what Lefebvre expresses as the problem where '[in order to] grasp a rhythm it is necessary to have be grasped by it; one must let oneself over, abandon oneself to its duration' (2013, p. 27). Here the intractable philosophical question of subject and object and their relations is posed in non-speculative terms, close to ideology, practice and the potential for resistance (Lefebvre, 2013, p. 36).

The representation of task-oriented conceptions of time projects an artificial nature onto the experience of time until it is absorbed into accepted models of temporal organisation which then become autonomic. That is, one becomes accustomed with engaging in activity x for duration y relative to a desired outcome. This becomes a performative, choreographed, routine.

> The relation between musical time and the body is required. Musical time resembles then but reassembles them. It makes a bouquet, a garland from the jumble. Through dance, first of all. Musical time does not cease to have a relation with the *physical*. If it begins with verbal rhythms, it is because the latter are part of the rhythms of the body. If it detaches itself from them, it is not in order to void itself but in order to reach all the so-called *physical* rhythms. Measure has this meaning: a means and not an end [...]. (Lefebvre, 2013, p. 64)

A regimented approach to space, time and productive bodies has an impact on the extent to which there can be continual improvement in human performance over time (Table 5.1). A rigid approach to time and motion, 'suffocates' (Adler, 1993, p. 97) the autonomy of the individual in the moment but does something different 'with both positive and negative effects' in the long term. The table below shows how rhythmanalysis can be built on and expressed in terms that make sense as a priority in self-tracking technology design. The next section looks at two examples.

Breath and Posture Tracking

I would prefer to think of the earlier example of distraction management not as a form of organisational learning but rather a relation between education, learning, training, or in the language of rhythmanaysis, 'dressage' (Lefebvre, 2013, p. 39). Although inseparable, in dressage, 'one breaks-in another human living being by making them repeat a certain act, a certain gesture or movement' (Lefebvre, 2013, p. 39). For example, a traditionally held view is that sitting up straight, or having good posture, indicates 'proper functioning of the body and contributes to good appearance' and that good posture promotes endurance and conveys poise, confidence and dignity' (Kendall & Kendall, 1968, p. 319). In an article published in *Physical Therapy* (1968) that also featured in Youth Physical Fitness, Kendal and Kendal suggested that Elements of a School-Centred Program be published as part of the President's Council on Physical Fitness offer the following among a set of 'Descriptive Comments that May be Used to Help Restore Good Alignment' such as: 'stand tall'; 'place weight evenly on both feet'; Keep both knees, not bent or pushed back'; ... and 'girls, standing in

Table 5.1. Future Oriented Self-Analysis.

	Past	Present	Future
Logic for mediating the body	Discrete moments	Events in real time, experiences and representations	Evidence-based construction of self in lifeworld yet to be lived
What is measured and analysed	Predetermined variables, pre-existing data	Real time data capture, multiple variables	Synthetic data
Supposed method of self-analysis	Reflective	In the moment	Predictive
Mechanism of self-correction (cognitive)	Applied learning	Behavioural change	Realising potential
Mechanism of self-correction (embodied)	Fixed, consolidated as memory and flattened	Embodied, goal-directed and reactive	Flexible, speculative, preventative and optimistic

a slouched manner with the hip protruding sideways gets you off to an early start on a middle-aged spread, so start avoiding it now" (1968, p. 326).

Breath control methods are another example of rhythmic bodily conditioning. These have been used for centuries in a variety of traditions to help people cope with stress, anxiety and anger. Pranayama are a set of yoga breathing practices used to consciously control and alter one's breath through specific patterns of inhalation, retention and exhalation timing. The general aim of rhythmic breath control in yoga is that it contributes to wellbeing by lowering stress and anxiety by activating the parasympathetic nervous system, via the Vagus nerve which can alter physiological parameters such as blood pressure and heart rate and when these are altered it can result in those using the technique to feel relaxed, energised and more alert.

A number of techniques have been developed which aim to train people into rhythmic breathing patterns. In the 1950s, Ukranian doctor Konstantin Buteyko (b. 1923) developed a series of breathing exercises. The technique which bears his name is designed to retrain the breath when one experiences hyperventilation or some degree of quickness of breath. The method works by having an individual undertake chronic repetitive breathing sequences of nasal-breathing and breath-holding. The physiological assumptions behind this method, which is not widely supported in Western clinical medicine, is that hyperventilation can lead to low carbon dioxide levels in the blood which disturbs blood pH and this in turn lowers the levels of oxygen in body tissue which has negative effects on the homeostatic state of the body. In retraining the body's breathing it is assumed that these effects can be reversed. When the method is used, people are encouraged to breathe through the nose. This protects the airways by warming, humidifying and filtering the air that is entering the lungs.

In the method, breathing is reduced by either altering breath rate or volume. The technique centres around a time interval referred to as the Control Pause (CP) which is the amount of time between inhalations that an individual can hold their breath. Once enough time has been spent practicing and the breathing pattern becomes instinctive the exercises are gradually phased out. Over time people who practice the Buteyko method are likely to see an increase in CP and a decrease in resting heart rate.

The Papworth Method, developed in the 1960s is similar but it places more of an emphasis on nasal breathing that engages the abdomen and diaphragm and not so much the chest. The Lamaze breathing technique is a psycho-prophylactic method used in pregnancy and labour as a way to manage contraction pain. The method is based on the assumption that controlled breathing can enhance relaxation and decrease the perception of pain. The method combines slow deep breathing, maintaining rhythmic breath cycles through the nose, mouth or a combination of both, keeping the eyes open or closed and focusing on a physical object in the room.

Similarly the 4-7-8 breath pattern technique involves a person inhaling for four seconds, holding their breath for seven seconds, and slowly exhaling for eight seconds and repeating this sequence for a given number of cycles, again with the aim of affecting the parasympathetic nervous system. The extent to which breathing control techniques are effective is also significantly influenced by posture. This is largely due to the way that posture effects the weight distribution of visceral organs and how this might impede movement of the diaphragm. Slumping while sitting can have a significant effect on diaphragm motility so maintaining an awareness of posture and breath can be an integral part of looking after one's body throughout the day.

There are a number of technologies now available to help measure posture and breathing such as pulse oximeters, tubular spirometers, opto-electronic plethysmography and respiration belts –all with certain limitations with regard to the extent they can provide data in real time, differentiate between different types of breathing or be discreet enough to use in every day settings. More recently devices including *Lumo Lift*, Upright Go and *Spire* have overcome some of these limitations and enjoyed some success in the consumer lifestyle technology market. One example from within this class of technology, *Prana* is a posture and breathing training device incorporating a wearable sensor that is worn at the waist and smartphone application. It tracks diaphragmatic breathing and posture in real time when discretely clipped to the user's waistband and connects to the user's smartphone to notify them as to when breathing and posture needs correction. The device also has a game-like function where the user can engage in quick two-minute training sessions called Tune-Ups to improve their 'seated health score' (prana.co). As the device's manufacturers claim:

> Our platform evaluates breath patterns, takes into account the effects of posture on breathing, and differentiates between diaphragmatic and chest breathing, three critical components of assessing the true quality of breathing, previously unaddressed by systems such as spirometers and pulse oximeters. Our algorithms are able to distinguish disentangle breath from posture ... utilising just one small sensor case worn near the belly or waist area. (prana.co)

The *Prana* device tracks the user in a combination of active and passive modes. In passive mode it tracks deep abdominal and chest breaths, respiration rate and posture angle and calculates a score for the user's breathing and posture. The device is pre-programmed with a library of breath patterns that the user can train against. In 'training' mode users can select a particular program –wake up, energy boost, focus, de-stress, wind down or sleep –which gives them a pre-programmed set of exercises to do in real time. In game mode, according to the device manufacturers:

> We designed *Prana* from the ground up to be an engaging and fun activity. In our active training mode, breathe your way to improved wellness by flying through our Zen Garden. Collect flowers by carefully controlling inhalations and exhalations, while maintaining good posture. You directly control the game character with your breathing, with the bird flying higher as you inhale, and lower as you exhale. Our library offers dozens of visualized breathing exercises for many distinct health benefits.

With *Prana* we see a technically mediated nudges towards constant improvement (conscious and unconscious learning) by alternating small refinements as one engages in repetitive actions, the confluence of cyclical and linear gestures (Lefebvre, 2013, p. 39). When Lefebvre notes: 'the analysis of the rhythms of dressage does not exhaust the understanding of social rhythms. Even if one inserts training for work, for the repetitive gestures of production, into the process of dressage ... more of less functionalised and linked to market conditions' (2004, p. 43).

This could mean that:

> [...] paradoxically beginning with bodies, the (concrete) universal that the political and philosophical mainstream targeted but did not reach, let alone realise: if rhythm consolidates its theoretical status, if it reveals itself as a valid concept for thought and as a support in practice, is it not this concrete universal that

> philosophical systems have lacked, that political organisations have forgotten, but which is lived, tested and touched in the sensible and corporeal? (Lefebvre, 2013, p. 45)

But this is only one way to look at it. When rhythm consolidates its theoretical status so to speak, it also introduces problematic analytical categorisations of the body. An example follows.

Industrial Athletes

StrongArm Technology are a start-up company based in Brooklyn that market a product called the FUSE Risk Management Platform, a wearable biometric harness coupled with a software analysis system for evaluating risk factors for workers developing lower back disorders resulting from poor lumbar positioning and movement in the workplace. StrongArm founder and CEO, Sean Patterson writes on a blog post featured on the company's website:

> I'm proud to say that over the past couple of years, we've developed our best tool yet to confront workplace injuries –the FUSE Risk Management Platform. We saw that continuous monitoring in driver telematics and industrial machine health had unlocked insights towards efficiency in the oldest industries in the world. We applied the same concepts to the most valuable thing in a corporation: human capital. In doing so we've been able to take the subjectivity out of injury analysis and conduct complete risk observations on every moment in a facility. (Strong Arm Technologies, 2019)

The harness translates sensor data (sensor orientation data) which includes lift rate, the flexion across the horizontal plane that exists between the left and the right side of the body, average twist velocity, maximum range of movement and maximum velocity (Patterson, & Kim et al., 2017) and calculates a score that represents the probability of injury risk to the wearer. In some framing text on the company's website, the 'industrial athlete' is introduced as such:

> We grew up seeing people like our fathers endure preventable physical pain on their bodies while providing for their family. 'Manual Labourers' or 'blue collar workers', unlike sporting athletes, have been overlooked in an age of rapidly evolving design and technology equipment. We didn't want to stand by and watch this injustice continue to impact the backbone of society, the lifeblood of our economy, the people who build the world around us. (Strong Arm Technologies, 2019)

While at the surface we might see the motivation for marketing the FUSE Risk Management Platform as an altruistic one, and one that is concerned for the overall wellbeing of their clients' labour force, the narrative surrounding this technology is unconvincing given the interest that insurers have in this type of data in reducing the number of insurance payouts they make. The data that the FUSE platform produces is also valuable because of its potential to be integrated with and linked to data produced by Labour Management Systems and Workforce Management Software that may already be in use in facilities that are adopting the FUSE platform. Labour Management Systems and Workforce Management Software take aim at employee engagement levels with pre-determined Engineered Labour Standards.

One of the vendors of this type of technology, Manhattan Associates, claims that the benefits of implementing labour management software are that it enables managers to identify and reward strong performers in the organisation and helps train employees who might not be meeting certain performance standards (manh.com/en-au/products/labour-management). The use of such systems offers managers seemingly objective and scientific bases for forecasting staffing requirements, assessing individual productivity and calculating, and awarding, incentive pay in a way that is not entirely fair. Engineered Labour Standards and self-tracking in the workplace overlook issues of bodily ability and neurodiversity. The technology is designed to be discriminatory and ableist. Its corrective and disciplining nature – warning workers if they bend too far or move too fast but doing this with a per-unit time productivity measure scripted into their logic –denies other factors at play which are not appropriate for measurement, documentation and reporting. For instance, when temporary workers are taken into facilities to perform physical tasks they are likely to be more inclined to engage in risk taking behaviour as to me measured as productive and, as a result of taking these risks, increasing their chances of being re-hired or offered permanent secure work.

Designer as Breeder

In the various circumstances where we might see this technology used there are naturally going to be resistance from users. Even with the convincing frame of worker safety and possibly the endorsements of workers' unions. Physical and mental tolerances are altered through training and at the same time bodily potentials are kept within a tightly calibrated range of externally defined upper and lower limits. This pushes the body to its limits but these are limits that are determined by decisions made during the design process. On the topic of translating a client's design brief, Ellen Crane, the product designer who worked on the FUSE harness writes:

> When designing for a client, your job is to show them the best possible outcomes, how to approach the issue or problem they are trying to solve. As a designer, how can you positively influence them? Be honest and genuine and create something that they may not have anticipated. Offer a unique perspective, a designer's voice. (Crane, n.d.)

It would be naïve to suggest that those who partake in self-tracking unquestioningly accept the authority or relevance of the datasets they, and others produce. Nor is it to say that those active in self-tracking communities give themselves over blindly to the logic of the device and how it produces data as it is not difficult to find evidence of both these issues being fiercely contested on various user forums online. Dawn Nafus and Jamie Sherman of Intel Labs offer some insight:

> The relationship between 'habit formation' and the limitations of devices is significant. On one hand, the habits/practices that most participants sought to instil in themselves generally (though not always) adhered to normative guidelines around health and good citizenship: exercise more, work more effectively, keep moods elevated, etc. On the other hand, these clearly are not passive consumers swallowing blindly the parameters of 'what's good for them'. In many ways they see their activities as a response to big data and big science dictums that make claims about the healthy body from on high. In the face of generalised, anonymous one-size-fits-all prescriptions derived from population studies, they seek to understand what is right for me. What is the optimal bedtime for me? Under what diet regime do I feel my best? What activities (sleep, caffeine, wheat, dairy, and other usual suspects) are particularly correlated with mood or energy in my life? (Nafus & Sherman, 2014)

Rhythmanalysis gives us a way to complicate the notion of designing self-tracking technology based on archetypal personas. As a commonly used method for designing technology in ways that are meant to be user-centric, archetypal personas give designers a set of qualities clustered into a finite set of user profiles. Bendelow and Williams (1998) describe several ways in which health oriented technology problematises the body and here we might use them to critique the reductive aspect of designing for archetypal users.

First, plasticity, bodies can be moulded, honed and refined through surgical and pharmaceutical and behavioural intervention. Second, bodies become bionic through the introduction of synthetic and autologous body parts, prosthetics artificial valves, pacemakers and a host of other devices. Third, bodies are becoming communal and interchangeable through organ donation, transplants and so on but also interchangeable in a sense that increasing information and data about the body makes them comparable and standardised across large populations. Fourth, and as an extension of the former point, bodies are also becoming interchangeable across species through xenotransplantation (cross-species transplantation) and the understandings of universality between species being produced through genomic databases and other advances in bioinformatics. Finally, the representation of bodies is becoming increasingly speculatively mediated gendered, and abstracted in time. As Barbra Adam reminds us

> …make[s] the taken-for-granted visible, to render explicit as well as question what is currently assumed natural. To this end [we] focus on the conflicts that arise within the industrial modes of life from a) the complexity and interpenetrations of rhythms: cosmic, natural and cultural; b) the imposition of industrial time on the rhythmicity and pace of ecosystems; and c) the prevailing emphasis on visible materiality and quantity at the expense of that which is hidden from view and latent. (Adam, 1998, p. 9)

But despite this, there has been a steady commitment by technologists to develop methods that apply a choreographed technocratic approach to managing time. In *Design for Dasein*, Thomas Wendt notes, 'ironically, this obsessive need launch and iterate is precisely what links design to phenomenology; the emphasis on praxis and time spend testing with users is a phenomenological concern' (2015, p. 9). But, the argument that the difference between design research and behavioural research is that the experimental techniques in the behavioural research reduce complexity by isolating variables to systematically discover truths about how a technology is experienced empirically. Whereas in design research, problems are addressed 'holistically, simultaneously and iteratively' (Forlizzi, Di Salvo, Bardzell, Koskinen, & Wensveen, 2011, p. 824) is not convincing with the commercial realties of industrial design. In theory, design research approaches the study the experience of technology use work holistically to highlight the interconnectedness of material, technical, social and experiential phenomena as literal, neutral and unbiased accounts. But this rarely translates to practice and the argument here can be put as such:

> Start-ups have fostered a large amount of backlash against the academic side of design research. Asking a […] start up founder to spend a month understanding who his end users are – what frustrates them, what their lives look like, why they would ever use his product – is like asking to throw away 30+ days of development time. Movements like […] Lean User Experience have done an excellent job of emphasising the importance of customer insight. And yet, they are easily bastardises into biased methods that end up reinforcing the shallow assumptions they set out to eradicate. On the other end of the spectrum, academic design research is often perceived as esoteric and disconnected from design practice. (Wendt, 2015, p. 9)

As David Rothenberg in his book *Hand's End: Technology and the Limits of Nature* points out:

> …the problem is that technological installation of human presence in the world invariably involves a series of choices. Choosing particular avenues of action necessarily closes others. As technical decisions are

> made, the original intention is fast channelled towards those possibilities which the technology admits. The more complex and encompassing the tool, the more it implies a unique, peculiar way of thought. This is why... successful programming of a computer leads us to imagine that our own mind works the same way. Clearly technique is not merely a means: just when we think the problem is solved, the machine reveals new troubles and possibilities alike. Technology wraps us up in its circle. This cycling path represents what it means to be human in all kinds contact with the external world based on the dream of rational, planned order. Does any single external plan guide the wish? Theodor Adorno writes that techniques can 'speak in a way which has nothing to do with the deliberate communication of a human message [...] What looks like reification is actually a groping for the latent language of things. (Rothenberg, 1995, xv)

Discourses about self-tracking materialise through interpretative processes of design. The next chapter covers this in some more detail.

Chapter 6

Anticipating the Quantised Self in Design

> [...] a lot of our devices sound like something, and if we were able to understand the rhythm of interaction with them as a piece of music, it would help us to design a more perfect union of man and machine. Although, this could perhaps be a crazy idea from too many late nights and too much caffeine, I feel very strongly that we need to listen to these rhythms. (Nudelman, 2016)

> Once [a technology is] created, [...] the opportunity for social choice diminishes. An implemented technology carries with it a powerful vision of the society in which it is to be used, replete with an equally powerful endowment of symbolic meaning and, sometimes, an obligatory plan for the way people will have to arrange themselves to use it. (Pfaffenberger, 1988, p. 16, as cited in Orlikowski & Gash, 1994, p. 197)

In this concluding chapter, I want to return to the idea that design has within it enabling and limiting forces that impact the agency of designers to do their work and influence decisions that translate to how the outcomes of design processes are experienced.

In places throughout the chapters I have pointed to designers' own reflections on the self-tracking systems they have worked on. Historical tracings featuring throughout the chapters have taken us through a number of cultural and technical trajectories which frame self-tracking technologies as we know them today. Some of these framings are more prevalent than others and the related-by-association aspect of this framing, as we've seen, is foregrounded in some instances (e.g. personalised health) and downplayed in others (electronic monitoring). What's common to the framing of the majority of self-tracking devices and systems that we see on the market is an articulation of the idea that these technologies are good for the user and using them helps bring about enriched and fulfilling physical and emotional experiences. In this sense self-tracking technology is for the same user at the same time, liberating and controlling. This contradiction is difficult to capture and represent when conducting research into how self-tracking technology is

The Politics and Possibilities of Self-Tracking Technology, 123–150

doi:10.1108/978-1-80043-338-020211007

experienced and what user needs, wants and concerns are. In what follows, I consider the way that discourses about self-tracking materialises allowing the logics and ideologies that we've seen so far in this book become embodied in ways that they produce positive and negative effects. When technologists and designers engage with self-tracking discourse their interpretations have the potential to either perpetuate problematic framings of self-tracking or to resist and even subvert them. As mediators of the lived experience of self-tracking designers and the work they do in conceptualising and iterating technology is political.

As a stage on which interventions into the negative outcomes of self-tracking could be established, the dialogue and the objects/tools that facilitate dialogue between the various stakeholders in self-tracking systems can be used to problematise positivist ideologies guiding and justifying decisions being made about the development of the technology. Not insignificant in this are the portrayal of archetypal users and patterns of use, and, how these are interpreted and translated as priorities in design and development.

As a coda to the book, this chapter considers the value of incorporating the 'quantised self' as a outcome of self-tracking technology use and as an idiosyncratic pattern of use arising out of the material architecture of devices and the logics of data capture, transfer and analysis that underpin these devices and systems as the modulate the messy realities of real user experiences. Here I want to invite the reader to think about how the visions of self-tracking can be incorporated into the broad set of approaches falling under the umbrella of 'design thinking' (Buchanan, 1992) and 'human-centred design'. This chapter describes and critiques aspects of technology design and places an emphasis on users and use-contexts, and, the relations of power between the various stakeholders in a given sociotechnical system. The way that I want to discuss users and patterns of use in this chapter has less to do with ideal users or archetypal people who might use self-tracking technology. My concern here has more to do with thinking about how we can account for users who are experiencing a state of mental, physical and emotional alignment – what I have referred to as quantisation – inscribed into the logic inscribed into the technology. I see this as an undesirable outcome of self-tracking but from some perspectives this might be seen as the goal. Discourses are formative in the way that they solidify and can be materialised in the process of iterative design. The outcomes of design research – interview data, participant observation, journey maps, personas – are generative in that they have a framing function but they are also open for interrogation from *different angles* during and after the design process. Different angles are important given frameworks that guide the design, administration and governance of self-tracking technology can easily fall out of pace with prescribed patterns of use.

Although my focus here is on design at the outset I will state that I am not a designer though I have worked professionally in design environments in other capacities and have extensively worked with designers on various projects involving the design of systems for data capture and sharing where responsible data stewardship has been a priority. I do not have any formal design training but as a researcher and an educator, I am informed and inspired by work that is happening in design theory and practice. Self-tracking technologies aren't like other digital devices and systems. The linked affordances of embodiment and data-capture situates them at the margins of a number of classes of technology. From the perspective of design, where a recognition of their politics should be a priority, self-tracking systems don't sit comfortably within methodologies that are based on designing for idealised and archetypal users, rapid iteration and bringing minimally viable products into their contexts of use with a view that they can be refined once out in the field. This process for bringing technologies to their potential users, which falls under the broad umbrella of approaches labelled 'agile' aim to sense changes in the market environment, including instances where regulatory environments are relaxed in ways that are permissive to the entry of new classes of data-garnering technology into an existing market segment. The so-called "agile" way of designing technology is geared towards being able to respond proficiently and flexibly to what are often shifting, ambiguous and unstable market environments with highly contingent and third-party dependent supply chains. The contingent nature of self-tracking technologies with data infrastructures and logics positioned outside the realm of moveable variables – that what data is captured and how it relates to the problem that a design is trying to solve is a constraining force on 'design thinking' and human/value centred design. These related approaches to aligning technology and its users seek to rationalise and draw out generalised principles of use that adds value to a users life. In theory this can be applied to the development of digital products and while these human/user centric approaches can and do increase the likelihood of better experiences of technology use, this is not to say that they assimilate so easily – and apolitically – to the design of self-tracking devices, systems and services.

Self-tracking futures and the technologies that are yet to come in this area are established by drawing on definitions, frames of reference, architectures decisions that are being all forged in the domains of policy, industry, market and design. Assumptions that construct self-trackers as rational, uncritical, consistent, or predictable downplays the complex relationships that users develop with the devices and the data they produce especially as this relationship changes over time. These questions rest on what can be assumed about design when it is thought of as a noun or a verb (Miettinen, 2008, p. 219) and how this shapes the instrumentality of design in shaping experiences of technology use (Table 6.1).

Design involves not only the configuration of technical components – like the sensors that make up a device or the straps that attach it to the body – but the meanings attached to them individually and as a whole (Franssen, 2008, p. 22). It is the latter, the whole, that gives self-tracking devices intentionality. It is too simple to think of the role of a designer as an actor that merely brings together

Table 6.1. Dialectics of Instrumentalisation.

Task/Function	Causal	Cultural	Operation	Technical Objective	Seeks to Normalise
Objectification (designing objects)	De-contextualisation, (commodity, use case, technical affordance)	Interpretation	Abstraction	Provision	Obligation towards efficiency and economy
	Reduction	Mediation	Translation	Layers (combine, stack)	Values and framings
	Association	Systematisation	Consistency	Adaptation	Inclusiveness
Subjectification (designing experiences)	Autonomy (sponsor, consumer, vendor, designer, user)	Representation	Feedback	Effects	Responsibility to sponsors and/or users
	Positioning	Imitative	Prescription, abduction, apprehension	Innovation	Progress
	Designing	Anticipation	Prescription, prediction	Specification	Order, efficiency, wellbeing, doing good

Source: Adapted from Khatchatourov et al. (2019, pp. 177 and 179).

technical morphological and aesthetic elements together in actionable or efficient combinations in a manner that has drawn on specialised knowledge and empirical evidence. In the development of self-tracking technologies, like many other intimate and embodied digital technologies, designers are faced great ontological complexity and must 'rationalise' a number of conflicting interests including their own.

Design methods and design discourses offer a way to drill through the dominant framing and are useful in sketching the contours of a productive mutual praxis that can get around some of the limitations within different disciplines such as HCI, STS and in the social sciences more generally.

The significance of the delicate relation between data, embodiment and politics is not easily perceived against the overall enthusiasm associated to self-tracking practices which are often given currency by perceptions that data collection and sharing will create unprecedented value and efficiencies in spheres *beyond* the individual life-world within which this data *makes most sense*.

I couldn't think of a better way to put it other than something the Uncertain Commons wrote in the context of activism in productive activities that involve collaborative intellectual labour – like design. To my mind, the quotation below frames perfectly the way that we should be thinking about the politics and *possibilities* of self-tracking as modality of automated surveillance, guided reflective practice and measurement, evaluation and not least of all representation in a way that at least accommodates its reversal or resistance.

> There is expectation, conjecture, and anticipation: modes of living that reorganise the dormant energies of the quotidian and eventualities that escape the imagination. We call these modes affirmative speculation. To speculate affirmatively is to produce futures while refusing the foreclosure of potentialities, to hold onto the spectrum of possibilities while remaining open to multiple futures whose context of actualisation can never fully be anticipated. (Uncertain Commons, 2013, as cited in Schifani, 2020, n.p.)

But what if we took future contexts of actualisation seriously as a way of facilitating productive exchange between those engaged in the critical study of self-tracking and those who design it? When introducing this book, I suggested that as a technical logic quantisation extends the mediation, alignment, error correction, smoothing out and removing of imperfection associated to datafication which is commonly observed when varying phenomena such as light, sound and shapes are rendered in digital form. I borrowed this term from audio production technology (e.g. drum machines, MIDI sequencers) where the 'quantise' feature is used to correct human variation and snap it into precise and predetermined grids. Similarly in image processing, quantisation is the lossy file compression where a range of values are flattened into one – for instance if the number of colours used to create an image are reduced or the brightness variation values are rounded then the image is represented in as an inexact approximation based on partial data. These techniques are irreversible and their effects are not easily perceptible to the observer. And by way of analogy when we apply the same logic of 'lossy' representation to how people are represented in the logic of self-tracking technology the same could be said about how perceptible these effects might be to the observer.

Representing Users in Design(s)

Scholars within Science and Technology Studies whose work I have already ended with in this book such as Akrich, Latour and Orlikowski and Gash have in their writing on the subject of understanding technology as inherently political offered various relational perspectives on power and agency. These analyses that either fall within or engage with Actor-Network Theory (ANT) show us how different actors involved in the design and assimilation of technology are embedded in relations and symmetries of power that aren't always configured to produce outcomes that are optimised to realise the best case scenario patterns of experience for users. They also show us that non-human actors such as discourses, prototypes and products exert some degree of influence in these relations.

Typical ANT inspired approaches to studying self-tracking system design would aim to show how various actors engage in work that seeks to ensure that the information guiding design processes and by extension the resulting systems and devices that result represent their interests. And in doing this actors engage in discursive, interpretative, creative and material labour that protects these interests within the interdisciplinary coalitions coming together to bring self-tracking systems, and logics, to the contexts where they are used and to the users who will be using them.

One can see this politics in technologies discussed throughout this book. The QS participants are exemplary for initiatives for capturing health data at the scale of populations but little concern is shown for those who are living with Electronic Monitoring technology under home imprisonment and house arrest when advocates speak out *for* more punitive and tighter forms of technology-enabled detention systems to be developed and trialed. Personal factors relating to someone's relationship to their workplace doesn't seem to be of much importance to the development sociometry and ergonomic factors and differences between people in terms of range of movement, pain thresholds and nociception are not accounted for in the data being fed to insurance companies via devices like the FUSE harness. But protocols of design go to great lengths to include users and I won't try to argue that the methods of including users in the process to digital technology design (participatory, co-design, user/human centric design, value-centred design) do not attempt to do this. What I will argue, however, is that the nature of self-tracking technology problematises these efforts. When archetypal people and situations inform design of self-tracking systems we need to ask which portrait should inform design for embodiment. Is it the person or is it the dataset? Can it be both? If it's a hybrid then what admixture and what is the overarching objective of how a portrait of the user is drawn on in the design process. Is it about supporting the design of humane technology or technology that's optimised for the most complete and accurate form of data capture possible within the technical and regulatory and budgetary limits of a particular project?

Either way, one can engage with personas in ways that foster empathy with users and empathy for the emotional impacts of self-tracking but this does not necessarily require empiricism. Fictional and speculative accounts of technology use be they artistic or from popular culture can productively inform self-tracking technology design (Pruitt & Grudin, 2003) and take us outside forms of representation that support the interests of those who are likely to benefit and profit from the uncritical design of self-tracking systems. Blending the empirical, archetypal and speculative, addresses the problem of representation in how we understand users in design, research, design-research and design/research. All three of these modes of "research" are ways of trying to understand what user representation and inclusion is in the context of production are relevant but with different approaches and to different ends. Embodiment of the sociotechnical, informational and political logics flowing through self-tracking technology then becomes a matter of, as Steve Woolgar (1991) puts it, 'configuring' (pp. 59–90) users for their enrolment into the task of validating design decisions by proxy rather than representing them meaningfully in the inscription of what can be done with a proposed technology and how that is done. In this sense, the voices of users are materialised in a discourse where those engaging in design-research speak on their behalf through rigid discursive forms such as 'empathy mapping' worksheets or elaborate graphical mappings of 'user journeys'. When these discourses materialise, they are drawn on in the design process support the 'inscription' (Akrich, 1992) of patterns of use into the technology. Although representing, configuring and inscribing patterns of user behaviours and patterns are techniques that I would argue are reductive – and to be fair often necessarily – in terms of how the capture and operationalise user experience there is something about these ways of constructing users and their experiences that can be repurposed. To meaningfully understand users requires some degree of interpretative freedom and anticipation. Designers, to at least some degree, must anticipate what might be done with a technology under certain pre-determined circumstances and from that a desired pattern of use or script is generated. As Richard Lester and Michael Piore (2004) note when design-research is seen as an enabler of *product* innovation compared to technical or social innovation. In the tenets of 'good' design:

> the customer has preexisting needs, and the job of the developer [or designer] is to identify those needs and then to create products

> that meet them in an optimal way. [...] In the interpretive view, the customer has no needs until they are articulated, and this articulation is what the interaction between designer and customer is all about. (p. 76)

The idea that participation and co-creation between designers, users and researchers leads to novel approaches to solving problems is not something that is going to be desirable in all cases. From the standpoint of actors who are sponsoring or administering self-tracking systems there are limits to human/user-centricity and trade-offs that will be made in regard to the extent to which a design proposal is aligned to *articulated* user wants and needs, *or* what we know is going to be in the best interest of the user of the needs of other more powerful and influential actors are put aside. The Electronic Monitoring technologies discussed in Chapter 3 are a good case in point. It is one example where we can appreciate that the designers intuition or desire to include users in the process is going to be at odds with the objectives of the technology. This could play out in two ways: (1) Include ex-users and discover how they developed workarounds and then develop stricter controls or (2) engage users in discussions about their frustrations with the technology and then use the information gleaned to explore how the technology could be made *more* punitive within the limits of normalised usability standards.

While the full spectrum of user experience cannot ever be captured in the process of inscription, the anticipatory nature of this thinking affords a wider type of representation and this doesn't have to be about users. It can be about scenarios of use or scenarios that fit a certain technological remit but use (i.e. doing something that creates data for a self-tracking system) is something that the user actively tries to avoid. If Human-Centred Design seeks to understand users' current problems and is oriented towards envisioning a future in which self-tracking offers a novel or alternative solution to that problem in a way that alters that reality based on a vision of the future where the present (and the problem) are static and don't evolve over time. If Human-Centred Design seeks to anticipate future problems based on an interrogation of users current reality then equally assumptions are made about users' lifeworlds. The question then is about the way that assumptions are justified and if this can be done in a way that is not too detailed or too superficial (Haddon & Kommonen, 2003). This is one of the issues with Human-Centred Design when it is applied to self-tracking systems is that one is dealing with a form of humanity that has undergone double abstraction, as persona or user and then as a body rendered in data and this brings us back to the question of if the intention of design is to act on the body or on the indicator which reflects an embodied process independently of its underlying causes?

Some self-tracking projects might fail their users because the nuance in the context of use may not be properly identified. For instance, in Li et al.'s (2010) five stage model for understanding user frustration with personal informatics technology suggests that users *prepare*, *collect*, *integrate*, *reflect* and *act* on data collected through devices and systems. It would be naïve to suggest that those who partake in self-tracking unquestioningly accept the authority or relevance of the datasets they, and others, produce. Nor is it to say that those active in self-tracking communities give themselves over blindly to the logic of the device and how it produces data as it is not difficult to find evidence of both these issues being fiercely contested on various user forums online. Within Li et al.'s model, barriers at the *preparation* stage are framed around the user being aware of what tools might be available and their overall satisfaction with the *informational* performance of that tool. The consequences of this, according to Li et al. (2010), are to do with the data discontinuity that might occur when one tool is abandoned for another (p. 561). At the *collection* stage, users observe and record different types of information about variables such as resting heart rate, physical activity, sleep, social interactions, behavioural patterns, thoughts and their immediate environment. Here the model considers barriers to be related to users motivation, time and effort in collecting data. One solution, according to the model would be to address this by adding incentives and reducing the amount of effort and time required to collect data using the tool. The *integration* stage is positioned between *collection* and *reflection* and is defined in terms of how much interpretative, cognitive and physical effort the user has to put into *taking* data from collection to reflection and problems with this stage, the model suggests, are to do with multiple inputs and formats, large volumes or inaccurate data at the point of collection (p. 561). What Li et al.'s model does not accommodate for is the extent to which this interpretative task is outsourced to technical processes and computational operations happening outside the users direct cognition and control. The *reflection* stage is given as users interacting with the outputs of the integration stage for instance with the data in visualised form. The *action* stage represents praxis that is meant to be based on newfound self-understanding, mostly tailoring their behaviour to meet pre-determined goals. A barrier within the model for this action stage, is that most self-tracking systems are not designed to *guide* users on what their next steps should be (p. 562).

As Li et al. (2010) argue the barriers identified at each stage 'cascade' (p. 562) where problems in one stage affect others or they are linearly 'iterative' (p. 163) and not flexible enough to support user informational needs as they evolve in complexity over time. But moreover, these authors recognise that lives are multifaceted and 'people have their home life, work life, daily interactions, physical activity' (p. 563) and so on. To overcome the barriers associated with each of the stages identified in their model Li et al. offer the designers a number of suggestions (pp. 565–566). They suggest that due to the cascading nature of barriers at each stage, they should consider systems as a whole so that they might accommodate users making alterations to the data they are collecting and allow for data to be transferred between devices and systems. Within each stage and facet of use, Li et al. note that there is an opportunity to either automating or de-automating processes that grant users more control and the decision about which of these would be more appropriate in a given scenario of use could come from research in the field of HCI or other related fields such as user observation and ethnography. But viewing points of user frustration in this way sees negative impacts of self-tracking technology use as acute instances that can be corrected within the stage-based linear logic of the model overlaid on to pre-conceived patterns of use. What is there to say about the more chronic impacts on users and how can be they be understood at the level of design?

When consultation is sought from prospective users this is still usually done with at least some of the intended purposes and desired outcomes pre-defined which then, to an extent, leads to certain predictions of technology usage work-flows and user journeys. Users invest in sociotechnical systems by putting time and information into them therefore, should return on this investment be measured through the cost-benefit paradigm where a technical investment is paid off through increases in functional efficiency or should it be considered through the qualitative impact on users lives both within and beyond the originally defined context of use?

This important question can be set aside when there is an over-emphasis on delivering a minimally viable or final product. 'the designer's responsibility to avoid the unforeseen and undesirable side and aftereffects of design' (Rittel, as cited in Protzen & Harris, 2010, p. 15). But going back to Li et al's model what is the nature of the incentive?

> One wants to know which goals to pursue; one realises that the political and ideological goals are too coarse [...] to yield instructions for concrete and far reaching decisions: historical philosophical programs less and less provide practicable norms for political decisions. (Protzen & Harris, 2010, p. 24)

However, these priorities can be integrated. They can be informed by representations of analogous technology in science fiction and popular culture. They can also be embedded with ancillary work that is directed by designers and researchers themselves where direction modelling technical trajectories can be based less on opportunism or lack of other preferred viable alternatives for what the sponsor of a technology might want but rather as due diligence and social responsibility in design. Part of the idea of an interpretative ethics involves:

> A product designer [attempting] to control the selection process, by controlling the environment in which his products operate. He may for instance attempt to require or encourage that a certain type of product is only used in pre-specified contexts or by pre-specified users. He may also attempt to alter the contexts of use in which products operate or alter the traits of users. He may for

> example offer training to users, or encourage such training, or he may recommend that adaptations are made to the environment in which the product is used. The designer's main ways of controlling the environment in which the product is used. The designer's main ways of controlling the environment include the authoring of manuals and direct communication with suppliers and users. (Brey, 2008, p. 73)

What is being described here is an operational discourse that can be shared among actors involved in the design and implementation of self-tracking technology. We could call this a form of social responsibility in the same vain as what many of the technology vendors in the self-tracking industry are doing to but that is *corporate* social responsibility and what I am describing here goes beyond that. The type of operational discourse (Protzen & Harris, 2010, pp. 7–8) or social responsibility that I am referring to here is one where designers have an understanding of the inner workings of the society in which they live and practice. This discourse *could and should* be developed through frames of reference that are amenable to incorporating ways of thinking about self-tracking technology beyond the standard conventions of a client/vendor/supplier/freelancer relation. But this is not just about apprehending the social dimensions of technology use and putting it into written and visual form. For it to be transformative in design, this enhanced discourse requires an acknowledgement of the inner workings of the economic and value producing mechanisms that are driving demand for the proliferation of self-tracking technology at both the consumer and enterprise level. There is a pressing need for the development of conceptual tools and heuristics that add structure to this discourse so that the role of designers and their position within the systems of value production and exchange they participate in can be understood as inherently dialectical. However there are two ways the this is curtailed (1) at the level of framing and accompanying discourses surrounding a technology or the proposed design for a self-tracking technology and (2) in the way that conflicts are rationalised relative to asymmetries of decision-making power over the production, iteration and administration of these technologies and their proposed designs. One way to think about this relation with a degree of attention to the conflicted position of designers is as Khatchatourov et al. (2019) argues, to consider:

> Rational criteria such as efficiency legitimate political and economic power in modern societies, and powerful institutions rely on technical disciplines [such as design] to satisfy those criteria in practice. This sets the stage for the confrontation of expert actors in many domains … These confrontations are not usually challenges to rationality as such but to the particular form it takes in specific cases. Critical constructivism argues that such events are a unifying theme in the study of the politics of [technology and sociotechnical systems]. (p. 62, also see: pp. 177–184)

An ethnographic study of factors relating to the design of applications to manage multiple chronic diseases conducted by Ancker et al. (2015) identifies four criteria that inform the design of self-tracking systems in a critical-constructive way. They are: (1) tracking this data feels like work for many patients, (2) personal medical data for individuals with chronic conditions are not simply objective facts, but instead provoke strong positive and negative emotions, value judgments, and diverse interpretations of data, (3) patients are motivated to track for different purposes, ranging from sense-making to self-management to reporting to the doctor, and (4) patients often notice that physicians trust technologically measured data such as laboratory reports over patients' self-tracked data. They conclude that:

> Developers of consumer health information technologies for data tracking (such as diet and exercise apps or blood glucose logs) often assume patients have unlimited enthusiasm for tracking their own health data via technology. However, our findings potentially explain relatively low adoption of consumer [health information technology], as they suggest that patients with multiple chronic illnesses consider it work to track their own data, that the data can be emotionally charged, and that they may perceive that providers do not welcome it. Similar themes have been found in some individual chronic diseases but appeared more complex because patients often encountered 'illness work' connected to multiple diseases simultaneously and frequently faced additional challenges from ageing or difficult co-morbidities such as chronic pain, depression, and anxiety. We suggest that to make a public health impact, consumer [health information technology] developers should engage creatively with these pragmatic and emotional issues to reach an audience that is broader than technologically sophisticated early adopters. Novel technologies are likely to be successful only if they clearly reduce patient inconvenience and burden, helping them to accomplish their 'illness work' more efficiently and effectively (2015).

Design-Research, Design/Research and Designer Subjectivity

> My hypothesis for why we are designing products at such a rapid rate is multifaceted. First, I should note that I am speaking within [a] context in which venture capital firms seem to be throwing money at any consumer product that comes along. This has created an environment of competition for funds much in the same way that corporate products compete for market share. The difference within

> start up economies is that instead of only launching 'on brand' products, there is large benefit of being fast/first to market. Ironically, this obsessive need launch and iterate is precisely what links design to phenomenology; the emphasis on praxis and time spend testing with users is a phenomenological concern. (Wendt, 2015, p. 9)

Defining uses for self-tracking technology usually begins with either a defined scenario of use such as 'how might we solve problem x using technology y?' or with open-ended propositions about how developments in one area of technology might be leveraged to improve or expand the scope of another type technology. Consider the following scenario. You have been commissioned to advise a multidisciplinary team on the development of a self-tracking system that integrates sociometry, electronic monitoring, tamper-proofing the sharing of data from self-experimental practices (prescribed by a third party for market/health/social research purposes) that incentivises users to send data to an online platform where it can be accessed under various terms by third parties. How do you explore the complexities and potential problems with this proposal in a way that would support the design of *something* resembling a workable, ethical system?

The critical analysis of the existing products or prototypes created to explore these possibilities eventuate out of the programs of design work that seeks to address the potential applications of technology as the next logical step in an established technical trajectory. Designers are engaged in the web of activity that follows once it has been established that the problem at the heart of such propositions is something that is worth solving and has an existing market or at least the potential to create a new market. Three ways of thinking about the focus, outcomes and power dynamics of design are given in Table 6.2.

Table 6.2. Perspectives on Design as a Process.

Perspective	Focus	Outcome	Location of Power
Traditional	Clients and/or sponsors	Instrumental	Micro-level negotiations between key actors
Constructivist	Other designers and actors engaged or interested in the design process	Political	Micro and meso-level structural interactions between actors with pre-existing power hierarchies
Critical	Broader society and industry, interdisciplinary dialog	Generative	Macro-level influence of tradition and culture on design processes in different sectors

Source: Adapted from Feng and Feenberg (2008, p. 111).

Self-tracking technologies tend to be imagined by drawing on conceptual and material resources that reflect one of these foci. In practice there will be multiple, often conflicting foci being contested through the design process.

Given this contingent nature of design, the alternatives explored are still based on desired applications of a certain technical feature or component, for example the affordances haptic feedback, tamper proofing, gamification, incentives to maintain use and share data or to make a device more ergonomic are point of departure for designers that seek *alteration* of an existing pattern of use, but, the expectations for the object being designed is *already* pre-established to some degree. The process of iteration does not always equate to a design be refined in a way that is properly human-centric.

When the *expectations* for a design alternative are challenged, what it is that is really being challenged is an expectation about efficacy within this pre-established vision or script for a technology and this has more to do with its function and efficacy for a certain purpose than it does to do with the expectations one might hold for its broader consequences within and beyond its predetermined scenario of use. For instance, designers have some degree of control over factors that influence the materiality and morphology of an object but less control over the relations that influence the functionality of the finished artefact and its patterns of use which then form the basis of future iterations (Ihde, 2008, p. 59).

When expressed as an objective for designers, the significance of negative emotions associated with intimacy of data collection that occurs with self-tracking technologies is not easily perceived against the overall enthusiasm for these practices as beneficial when considered in a purely rational sense. Benefits are often given currency by perceptions that data collection and sharing will create unprecedented value and efficiencies in spheres *beyond* the individual life-world within which this data *makes most sense in terms of what's being captured as much as how it is being captured.* Designers can contribute significantly to representing the emotional dimensions of self-tracking technology throughout the design and production process. In an interview with *Dezeen*, Fitbit's Gadi Amit offers his opinion on this. He states:

> So I think the biggest problem with this industry is that it has focussed too much on IQ. It doesn't focus much on the EQ [emotional intelligence], on the emotional quality of the design and the technology. And that is the biggest challenge. And this actually where designers excel, rather than technologists […] And then we go back to the whole issue of introvert/extrovert. A lot of people are trying to leverage data information for businesses. This is in direct contrast to a lot of perceptions we have about our own right to have control over our own personalities. And the industry is far from being mature about that, far from being mature. Some of the issues we have with Google Glass for instance now in San Francisco, there were nearly fist fights over Google Glass coming into bars where people tend to be more intimate. They don't want to be exposed to the whole social web. It's a major error on the

> part of people who never thought about the social and emotional state of people. So design has a lot to do with it and I think we will have maturity growing in this industry whether the industry likes it or not. (Fairs, 2014)

One consultancy that try to embody the ethic of incorporating the emotional experience into the already complex relations surrounding the problems that a technology is trying to solve are native Silicon Valley firm, IDEO who regularly hosted QS meetups (see: https://quantifiedself.com/blog/tag/ideo/). IDEO's founders had a background in industrial design and approached production with an emphasis on interaction between they objects they designed and their users. The focus on 'human factors' in the interaction between humans and technology positioned IDEO well in Silicon Valley and among the corporations and start-ups that were developing products and services that bridged the physical and digital worlds (Gram, 2019).

The 'human-centred', 'design thinking' approach championed by IDEO and widely adopted by others involves a five-stage process that is non-linear and adaptable for different purposes. At the centre of the design thinking process is the notion of human, or user, centricity. The goal of using the design thinking method is to 'create attractive products that are highly useable and offer desirable functionality' and that the result of using design thinking is that products (and services) are designed with adequate exploration of technical possibility and connecting this to the wants, needs and desires of users (Spreenberg, Solomon & Joe, 1995, p. 165). The IDEO five stage methodology is described in Table 6.3.

But, these methodologies are deeply flawed on the basis of the way they construct the 'human' within its relations to the technical objects and logics of datafication. Consider how success is defined in this context. In an article about self-tracking by Vanessa Monogioudis, a design researcher at IDEO defines it through the concept of 'stickiness' which is a term used to describe the extent to which a digital product or service resonates with users and encourages continued use. Monogioudis writes:

> These technologies offer huge potential [...] Self-tracking can help people better understand their body and what affects it, to make more informed choices about their health. Despite this promise, self-tracking services [...] don't reach people who'd benefit the most [...] And they don't always deliver meaningful data. Users are left with numbers but little insight [...] Many self-trackers stop after an initial period of curiosity. (2015)

For Monogioudis and her colleague at IDEO, Hana Peres, the way to design better and more human centric self-tracking technology is facilitate the sustained conversion of data to insight and insight to action. They cite the example of Mira a fitness tracking device that IDEO worked on. In Monogioudis' words Mira is a

Table 6.3. Annotated IDEO Methodology for Human-Centred Design.

Stage	Activity	Outcome
Understand	A 'kick off' workshop and discovery session to understand concerns, biases, organisational structures, and expectations on both sides. Desk research and review of client-side materials relating to products and users.	Common language between all parties that aims to operate above lowest common denominator terms of reference
Observe	Users are observed with the technology in familiar surroundings so that the technology is viewed in environments which are, or, mirror actual contexts of use. Observations are usually video recorded.	Supplementary data on user experience seemingly free from vendor-related assumptions and biases.
Visualise and Predict	Information from the first to stages is used to develop hypothetical and archetypal personas and uses cases.	Characterisations and portrayals of how the technology is used in by these persons in these contexts are used to make predictions about problems and opportunities.
Evaluate and Refine	Certain scenarios and designs resulting from the work in stages 1–3 are selected and mock prototypes, usually paper based are built and tested with users. The models used in this stage will often go through several iterations.	Open prototypes
Implement	The end product of this process can take the form of a series of blueprints, process diagrams, design guidelines or working software applications.	Minimal viable product

product 'designed for busy women, the device aims to integrate more activity into their daily routines. First, by getting to know what exercise they like – say, evening power walks – which they enter manually' (2015). The 'boosts' feature produces 'stickiness' by offering users contextual incentives such as 'the incline button on the treadmill is there for a reason, and that reason is to help you get an amazing body' and 'Walk as fast as you can during your next walk – like when your hungry and trying to beat a stranger and trying to beat a stranger to the hostess stand at your favourite restaurant' (Monogioudis, 2015). But making technology more human-centric by trying to add context in order to focus and motivate users towards goals is problematic in terms of its projection of norms and the ability of contextual cues to translate across class, gender, bodily ability, culture and an almost unlimited array of intersecting human factors which can add to the alienation felt by users in ordinary everyday circumstances.

Australian designer and researcher, Leah Heiss shows us a different approach to achieving genuine human centricity in the design of self-tracking technology. emotional experience of self-tracking as a central focus of human centred design. For Heiss, wearable, self-tracking technology is something that one *lives with* and it impacts, especially when we are considering medical self-tracking devices, one's sense of dignity. The objects Heiss designs – hearing aids, bio-sensing emergency jewellery, jewellery that for administering insulin through the skin – are beautiful objects in their own right and they provoke us into thinking about functionality as something that has a double meaning. In an interview with *Impact Boom* Heiss states:

> With the designer not at the table, it really means that those things just go through and, by the end of the design process, we have a technology that doesn't necessarily gel with its user group [...] Prototypes are more than just material artefacts. We think of them as boundary objects between disciplines and they enabled people to come together and make decisions and move on, and get over things [...] We come together around the prototype and also the prototypes are really powerful for working with users. So what we do, is we share the prototypes with users in focus groups and they try it on, they put in their handbags, they put it in their pockets, and I've observed that and they emotionally get really involved and invested in the design process through those prototypes [...] So rather than, yes, you've got the diagnostic, we know how well you are, or we can understand your disease inside out in terms of numerical data. But how do you then move forward into the technology of the lived experience and ensure that it's not impacting that daily routine in that sense? The difficult stuff is the embarrassment, shame, stigma, difficulty, isolation, awkwardness [...] So we talked around the prototype. I think of them like a campfire. And if you can really spend time

> listening and absorbing and understanding that information and then taking that into account in the design process, then you build a great reservoir of respect. (Stephens, 2018)

Speculative Objects

It is tempting to think of digital technologies with the capacity to mediate the body as an innovation which overcomes the inaccuracies associated with prior forms of recording and measurement. But there are important lessons from attempts made throughout history to mediate bodily processes. Some of the first attempts to use machines to represent the inner workings of bodies were mechanical automata. These objects were attempts at simulating biological processes so that the ways that bodies worked could be demonstrated to others and used to produce more scientific understandings of human physiology. By imitating life by technical means automata made it possible for people to develop forms of self-knowledge that were made meaningful by analogy and simulation. For instance, there is the work of French inventor and artist, Jacques de Vaucanson (b. 1709) who in the 1720s after leaving the clergy moved to Paris to study music, mechanics and anatomy began to conduct various experiments in mechanising mammalian physiology. His most famous invention was an artificial brass duck designed to ingest, 'digest' and excrete grain and water. The duck was produced at a realistic scale in simplified form with small openings in its casing which allowed viewers to observe the process of digestion. The duck made its first public appearance in 1738 in a rented hall and in the years that followed, de Vaucanon's practice of building automata garnered him a reputation that resulted in being appointed to the Paris Academy of Sciences as an 'associated mechanician' – a qualifier that was affixed to this title so that it would differentiate him from the scientific fellows at the academy on account of the institution's aversion to engaging in commercial work (Riskin, 2003, p. 601).

The duck is an interesting object to consider in the analysis of self-tracking technology because, like contemporary examples of technology that mediates the body, it too was developed from other preexisting technologies. In de Vaucanson's case, the automata were developed by taking inspiration from self-moving machines that had been around since antiquity. The duck was also an educational technology, a demonstration of technical innovation of the times and an amusement which the public would pay to see. As Jessica Riskin (2003) notes,

> ...their value as amusements lay principally in their dramatisation of a philosophical problem that preoccupied workers, philosophers and kings; the problem of whether human and animal functions were essentially mechanical. (p. 601)

The reproduction of mammalian physiology in machine-form provided an important allegorical model that was used in communicating advances being made in the anatomical sciences beyond the discourses of the medical and intellectual elite. As Riskin (2003) describes:

> By the late eighteenth century, automata were imitative internally as well as externally, in process and substance as well as appearance. Cartesian dualism, which had exempted consciousness from mechanist reduction ... which allowed for an infinity of possible mechanisms underlying nature's visible behaviours gave way to an emergent materialism and to a growing confidence, derived from ever-improving instruments, that experimentation could reveal nature's actual design. These developments brought a new literalism to automata and a deepening of the project [of mechanically mediating the body]. The designers now strove, not only to mimic the outward manifestations of life but also to follow as closely as possible the mechanisms that produced these manifestations. (pp. 602–603)

However, automata were not intended to perfectly emulate the function of physiological processes, they simply imitated a predefined series of steps that *mirrored* certain physiological functions, de Vaucanson wrote in a letter to the *Abbé De Fontaine*:

> I don't pretend to give this perfect digestion, capable of producing blood and nutritive particles for the support of the animal. I hope nobody would be so unkind as to upbraid me with pretending to be any such thing. I only pretend to imitate the mechanism of that action in three things, viz. First, to swallow the corn; secondly; to macerate or dissolve it; thirdly, to make it come out sensible changed from what it was (1742/trans. 1979).

Mechanical automata have much to teach us about the artifice of technologies that mediate the body. This is not merely a question of the fidelity with which mechanical systems could mediate and represent physiological processes, but also one of interpreting beliefs of how bodies function and how emerging technologies might be taken advantage of in order to foster and encourage reflection on the body-machine analogy present within the logic of self-tracking technology, particularly with regard to the technical aspects of self-tracking technology and the opacity of design processes. For example in instances where automata take on a life of their own. This tendency is modeled in *Game of Life* is a cellular automaton designed as a single-player game by British Mathematician, John Conway. Played on a two-dimensional infinite orthogonal grid made up of square cells each of which exhibits a binary state of being either 'alive' or 'dead'. Each cell interacts with eight of the cells surrounding it horizontally, vertically and diagonally and at each progression of unit time: (1) any cell with fewer than two live neighbours dies by underpopulation; (2) any living cell with more than three live neighbours dies by overpopulation; (4) any dead cell with exactly three live cells becomes alive again by reproduction. The game, which simulates real-world processes, particularly of *emergence* but given its tight rule-based nature and that the patterns that emerge can be classified into types according to how they behave, their representation of self-organisation and free will within the ontology of the game is partial.

This partial free will is one way to think about how self-tracking devices and systems produce the state of quantisation. If we deliberately foreground this state as an outcome scripted into the technology and engage with it speculatively in interdisciplinary dialog in the way that Heiss does with her prototypes then designs in their own right might be used to raise questions about their own politics.

On this point I am reminded of a passage in Bruno Munari's book, *Design as Art* (1966) where the Italian inventor, artist and designer reflects on a series of objects he made between 1933 and 1954. These objects, which he termed *machine inutili* (useless machines) were simple constrictions of cardboard, metal, wooden rods, thread and glass. Munari (1966) wrote that, 'with the slightest movement of air, these simple constructions could be set into motion' (p. 15). But when he talks of these useless objects being set into motion Munari is not just speaking literally as to suggest that the object is akin to a wind chime or hanging mobile, he's suggestion other lines of interpretation. On the one hand, these objects were useless because unlike other 'machines' of the age they did not produce goods for material consumption, nor did they create efficiencies in labour or time that resulted in the production of surplus value or capital in some form (Munari, 1966, p. 23). On the other, they were considered, by some, to produce goods of 'a spiritual kind' which had the affordance of 'cultivating' discussions about the form and function of objects in a way that could create a counter narrative to the technophile rhetoric and hubris of industrialism.

Much of the design discourse around self-tracking inextricably presents a radical vision of a new world: one in which technologies attached to the body bring about a transformation where knowledge about oneself is used by individuals or groups in their everyday lives without negative consequences. But where are the counter narratives without being written off as panicked and too dystopian to be taken seriously by designers? In order to incorporate such narratives into the design language of self-tracking technology it is necessary then to reframe contexts and experiences of use allowing for *unanticipated* affective residues to be inserted into the materialist analysis of self-tracking scoiotechnicity (Hickey-Moody, 2020). The attempt to re-frame self-tracking technology in this way takes us beyond the morality of self-tracking technology and is one of the more subtle moral and ethical aspects of design, dissemination and advocacy of self-tracking technology.

The Unfit Bits project has arisen as a response to the increased privatisation of self-tracking data. Led by researchers at New York's Data and Society institute, Surya Mattu and Tega Brain. The project explores spoofing techniques with the devices to create false data to in order to allow users to qualify for health insurance discounts without having to share their actual private data. After interrogating the workings of health-tracking devices a series of simple techniques and crude spoofing devices have been developed. For example, mounting a self-tracking device to metronomes, pendulums or drills will produce an artificial data set for step counts and physical activity levels. Through the Unfit Bit website and social media channels users also contribute their own techniques for spoofing the devices and critical information about data privacy, and the self-tracking industry is circulated.

The *Unfitbits* project is not an intervention that is based on re-iterating a device but rather it offers a way of thinking about resisting the logic of self-tracking on the basis of re-contextualising standard uses and keeping key goals of the technology intact which is crucial if there is to be legitimate critique that resonates across a design and development team. In the obtuse patterns of use that the project depicts the Fitbit is still measuring the same unit, a step, but the input that results in data being recorded as a step, presumably being taken by a human, arises through a completely reorganised assemblage of kinetic relations between objects. The device produces embodied data though it is completely disembodied in terms of its physical relation to the subject of the data that it is meant to be collecting. *Unfitbits* challenges individual responsibilisation in the care of oneself and one's health and provokes thinking about individual agency over the devices logic and materiality first, and then considers the potential for collective action once the extent to which the devices inscribed logic can be subverted is understood. When data capture is a priority in the design of a sociotechnical system, its logic has a derived intentionality (see Table 6.3) where:

> [a] designer or maker has the same relation to technical artefacts as a breeder has to the animals he breeds. The designer attempts to create a certain artefact with desired properties but is not in full control of the outcome. Concrete artefacts are a compromise between the designer's ideals and the contingencies of the physical and social world in which the designer operates. While a designer is not fully in control of the outcome of his designing activity, he is even less in control of the success of his artefact once let loose on the environment i.e. the marketplace and the world of users [...] it is the intentions and choices of users, regulators and others, as well as random events, that determine whether it [is successful]. (Brey, 2008, p. 72)

But this type of thinking, that have little control over the outcomes of their designs, dis-acknowledges the fact that there are ways that designers *do* (and should) play a role in the 'future mediating roles of technology in-design' (Verbeek, 2008, p. 99) and that there are ways that designers can bring certain concerns to the table and present them in a manner that could make them less likely to be ignored or relegated to the margins through iterative processes of product rationalisation.

In practice, though difficult, the politics of self-tracking needs to be considered through certain values so that when designers perform the task of choosing which patterns of use designs and iterated on, alternatives are given credence (see Fig. 6.1).

As Verbeek (2008) argues:

> Design specifications should be derived from a product's intended function and from an informed prediction of the product's mediating roles and a moral assessment of these roles. A key tool to bring about this coupling of design context and use context, however trivial it might sound is moral imagination. A designer can include the product's mediating role in his or her moral assessment during the design phase by trying to imagine the ways the technology-in-design could be used by shaping user operations and interpretations from that perspective. Performing a mediation analysis can form a good basis for doing this. It cannot be guaranteed that designers will be able to anticipate all relevant mediations in this way, but it is the maximum that designers can do to take responsibility for the mediating role of their products. (p. 10)

We see an example of this type of anticipation in Adrian Norman's *Project Goldfish* (1978) which was a mock feasibility study for a system to gather, record and sell sensitive data written in the form of a business proposal. The proposal, which is written from the point of view of a consultancy which has been commissioned by multi-national corporation to explore the feasibility of such a system outlines in detail what the market, structure, and technical and legal barriers for the system – which is modelled on offshore tax havens but for data rather than money – would be. Interestingly, the proposal considers potential negative impacts. Norman (1978):

> If Goldfish can provide the kind of privacy and security of information that Swiss banks have achieved for financial transactions, then it will attract, like those banks, the business of

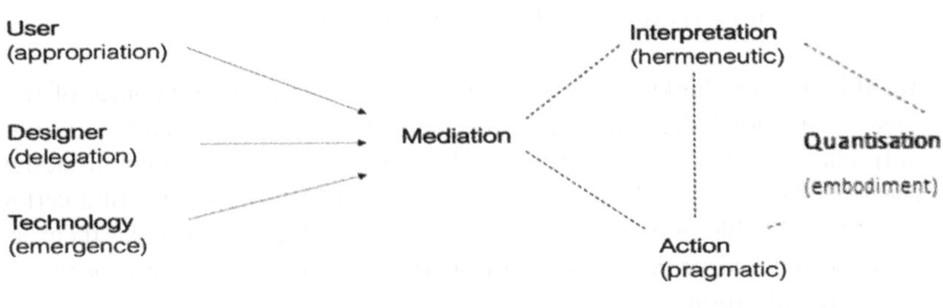

Fig. 6.1. Model of the Design Process.

> leaders of potentially unstable states. The Swiss bank secrecy laws were established for the best of reasons and were further strengthened to provide added protection for Jewish victims of Nazi persecution. But measures to protect the innocent can also be cloaks for the guilty, as courts all over the world frequently discover [...] This kind of facility should also appeal to law enforcement and related agencies who feel excessively constrained by data protection legislation in stable democratic states facing subversion [through terrorism or extremist political opposition]. Their intelligence files could masquerade as statistical systems, and nobody would be any the wiser. (p. 9)

What is also interesting about the *Project Goldfish* proposal is the extent to which its hypothetical nature allows for real world considerations to be explored as if a data haven was a real prospect. On the design element of his proposal, Norman (1978) notes:

> The terminals attached to Goldfish must be standard devices so that they do not attract special attention from authorities in countries unsympathetic to the system [...] Authorised users will identify themselves to the terminal by signature, voice pattern, hand size or other physical attribute. The terminal will identify itself to the central processor by its build in unique answer back code at sign on and will be disconnected and reconnected by call back when the user-terminal combination is approved. Such terminals are likely to be widely available in the near future because of the requirements of private operation over public networks with one terminal having access to many different services. (p. 11)

It is astounding to think that within Norman's tongue in cheek but incisive proposal we see the outline of so many contemporary trends in technology – biometrics, data-driven intelligence, two factor authentication and the blockchain – functioning cohesively as a model of the type of data garnering systems imagined in many of the self-tracking systems already discussed. Speculative modelling of technical systems does important conceptual work in contributing to the advancement of a critical agenda in the design of self-tracking technologies. Speculative design arises out of radical artistic practice, within the humanities and social sciences and within field of HCI where it takes the form of 'design proposals or built artefacts [that] are used to imagine alternative sociotechnical configurations of the world as to interrogate questions about values and politics through design' (Wong & Khovanskaya, 2018, p. 176). The term 'critical design' was introduced by Tony Dunne and Fiona Raby as a response to design paradigms that supported the status quo and affirmed dominant views of markets, technical utility, and human relationships with technical objects. Against such 'affirmative design' methods speculation was viewed as a way to dialectically challenge the notion of design outcomes as being 'post optimal', that is less concerned with efficiency, usability, quality, or durability and rather being driven by market forces which rushes products to market.

It has a history of use in the field of Human-Computer Interaction (HCI), and since having emerged within that literature in the mid-2000s it is presented as a practice linked radical art and critical theory (Wong & Khovanskaya, 2018, p. 175). Yet, as Richmond Wong and Vera Khovanskaya have argued that given that speculative design is a practice that has been introduced into HCI via the work of Xerox PARC's RED group, that it 'works … because it is highly compatible with other forms of corporate speculation [such as scenario planning]' (p. 175). Further, given that the critical focus with speculative design is re-oriented in production environments where there are multiple competing interests, the outcomes of speculation and its critical dimensions are likely to reify rather than challenge dominant practices. With this in mind my focus in this chapter is to not only discuss ways in which a realignment of speculative design practices could occur but to also show how this can, potentially, be done within dominant ideation and design practices.

For Dunne and Raby (2001) critical design creates a space for the cultivation of conceptualisations and hypothetical designs for technical objects from a point of view that is outside the constraints of the marketplace. Although critical design was originally proposed to be something that was more suited to academic research and teaching it is being adopted to a degree or at least mimicked in various types of corporate and commercial 'design innovation'. Like the notion of critical design, speculative design uses the proposal of objects or systems to open up and explore possible scenarios and futures might arise from using a given technology in a given context including generating discussions about which of those scenarios are *preferable*. Hypothetical proposals allow for this exact type of anticipatory thinking to be taken further into the production process and through to final products.

When used as probes which according to Gaver et al. (1999) serve interdisciplinary design by facilitating the capture of information about a given problem and its proposed technical solution and to challenge rationalist methods and explore designs that would elicit emotional and creative responses rather than objective responses which are read relative to a pre-determined and expected patterning or 'script' (Akrich, 1992). As Gaver et al. (1999) write:

> We approach research into new technologies from the traditions of artist–designers rather than the more typical science-and engineering-based approaches. Unlike much research, we don't emphasise precise analyses or carefully controlled methodologies; instead, we concentrate on aesthetic control, the cultural implications of our designs, and ways to open new spaces for design. Scientific theories may be one source of inspiration for us, but so are more informal analyses, chance observations, the popular press, and other such 'unscientific' sources.' (p. 24)

Speculative probes have three main functions in the design of self-tracking systems (see De Souza, Butt, Jethani, & Marmo, 2021). Participants and researchers to externalise their understanding of the themes being explored. Rather than relying on more extractive methods (like a survey or an interview), it allowed participants and researchers to collaboratively build, shape and check their understandings with each other through a progressive interrogation of the object. Working with a speculative object allows exploration of ideal scenarios of use, while also allowing space for people to view prototypes as confronting or threatening. Secondly, they can take pressure off discursive modes of communication by allowing participants and researchers to draw on their respective understanding of the metaphors inherent in a speculative object which is intended to be a proxy – one that is open to a wider range of possibilities – for an intended design. Thirdly, they function as augmentations to other sets of expressions and knowledge practices. People were able to position the probes in relation to their bodies; proximity and distance became powerful metaphors themselves for all kinds of thematic explorations.

It is Annemarie Mol who points us in the direction of a way of thinking about ontology that can inform human-centric approaches to self-tracking that might facilitate deeper critiques of their logic. Mol argues ‘ontology is not given in the order of things, but that, instead, ontologies are brought into being, sustained of allowed to wither away in common, day-to-day, socio-material practices’ (Mol, 2002).

Table 6.4 shows us why is it is difficult to offer a prognosis for self-tracking through a lens established within the disciplinary purview of the humanities and social sciences despite the great deal of critically oriented work happening around self-tracking. Researchers do not sit outside this relation. The conflict of interests then is twofold: (1) at the level of framing and accompanying discourses surrounding a technology or the proposed design for a self-tracking technology and (2) in the way that conflicts are rationalised relative to asymmetries of decision-making power over the research that guides production, iteration and administration of these technologies and their proposed designs even though designers do this work – and do it well – as part of the discovery process at the outset of a project. As Pantzar and Ruckenstein (2015) state:

> Techno-anthropological projects need to intervene here in order to ask whether this is a valid goal: in the long run, research with no promises of immediate financial productivity might be socially and economically more rewarding simply because it aims for a deeper understanding of long-term effects and trajectories, rather than simply taking for granted the current state of affairs. Consequently, reflexive and critical accounts of researchers' involvements with technology companies, including the developers and promoters of their wares, are of great interest for techno-anthropological research, because they can illuminate the commercial, political, and ethical underpinnings of research

Table 6.4. An Ontology of Self-tracking.

Emotional Factors	Social Factors	Ergonomic Factors	Technical Factors
Perception of value, change over time, cognitive effort	Influence of communities, peers, subcultures and institutions of power	Physical effort, comfort	Technical limits (performance, processing, storage and analysis)
Motivations (intrinsic and extrinsic, commitment, pride, sense of achievement)	Narratives, histories, biographies (of users and objects)	Level of integration into everyday technology use	Rules, norms and standards regarding safety and performance
Memory and cognition (learning); mind/body relation	Representations of self-tracking in popular culture	Physical adaptation and effort (training)	Transparency around processes of data capture and interpretation
Disorder, frustration and confusion, shame, embarrassment	Incentives to use the technology	Aesthetics	Accumulation and transfer of technical capacities across different classes of technology

> endeavours and promote down-to-earth perspectives on how technology companies – in the course of affecting and shaping people's daily lives and futures – interpret, design, and aim to produce those lives. (p. 14)

But it is not easy to do this empirically without some degree of speculation. One issue in particular is that the difficulty that researchers, and consumers, have to subject powerful industry players to adequate scrutiny given that much of the nuance of their practices are hidden from view by mechanisms that protect proprietary or commercially sensitive information or personnel working in this area are often blocked from access to researchers or limited in their ability to contribute to ethnographic inquiry by way of non-disclosure agreements or lack of trust between practitioners and academic researchers even when stringent protection measures are put in place.

The conflation of design research and research into processes of design has consequences for the extent to which design practitioners can be reflexively critical about their practice, or on the extent to which designers might translate their own critical reflections on their practice into productive outcomes for the work they do in self-tracking technology design. Design research exists to simplify complexity and generate evidence so that prediction or assumption can be limited within the decision making that does into designing for users (and uses). For instance, when designers use archetypal personas to develop models for understanding user behaviour they translate and reify external forms of power into what are assumed to be neutral points of departure from which ideas are generated around a human user at its centre. The design of self-tracking technology proceeds from the abstract to the concrete but only to talk back to the original abstraction as a proxy of a human user embodying the outcome of design.

For example, in *Raw Data is an Oxymoron* (2013), Lisa Gitelman (2013) notes that 'data are familiarly "collected", "entered", "compiled", "stored", "processed", "mined" and "interpreted". Less obvious are the ways interpretation haunts its predecessors' (p. 3). And the discussion of data-provenance given in Chapter 2 brings some of this to light through frameworks that could, if given genuine credence, assimilate into, and make a significant contribution to the inscription of logics of information flow and their embodiment, and, the mapping of user-experience at the stages of technology use where data-outputs are interpreted and acted on.

Human centred design and design thinking are premised on an assumption that collaborative work is valued for its ability to not only deliver efficiencies but to also unlock the under articulated politics associated to various modes of production such as agile design and its methods of rapid prototyping in units of time referred to as sprints, and iterating towards a minimal viable product which is then refined. There is a hopeful speculation that engaging in co-design and design thinking will increase the impact and value of devices and systems in terms of how they positively influence the quality of life for the people using them. Especially for connecting with the original emancipatory intent of these design and research practices (Teli, Di Fiore, & D'Andrea, 2017).

> The plan is an argument, reflecting the deliberations of designers and their efforts to integrate knowledge in new ways, suited to specific circumstances and need. … The power of design as deliberation and argument lies in overcoming the limitations of mere verbal or symbolic argument the separation of words and things, or theory and practice that remains a source of disruption and confusion in contemporary culture. Argument in design thinking moves toward the concrete interplay and interconnection of signs, things, actions, and thoughts. (Buchanan, 1992, pp. 19–20)

Design as a medium in its own right; the quantised self/body as a medium in its own right making epistemologically (in terms of learning) and practically (in terms of production methodology) design thinking and 'medium thinking'. A move away from thinking about the use of self-tracking technology as something that happens in an environment in which bodies are present and activated for data capture towards the technology as embodied and here Bratton's notion of 'the stack' which was discussed in Chapter 2.

As Keller Easterling (2018) notes, 'design that has any hope of effecting change manipulates the organisation [of production] as well as the narrative that attends it'. This is a shift from design thinking and human centred design to what Easterling – following McLuhan who argued that any understanding of the social and cultural change associated to the use of technology required an understanding of how media function as environments – terms 'medium thinking' which considers the 'matrix of activities that inflects' (2018) and inscribes the affordances and capacities which bring about experiences of self-tracking as ingrained cultural practices that command acknowledgement during design and then retrospectively. The perspective on self-tracking that we would need, which cannot fully cohere within design thinking requires necessary critical moments to occur in the process that stakeholders undertake in apprehending users and contexts of use. These moments rarely occur and therefore design falls back to lowest common denominator accounts of use and a state where components of knowledge, might well be called, by analogy a 'unitary theory' of human centred design falls short of its goals. The aim of this chapter has been is to sketch out some of the elements of a shared discourse between 'fields which are apprehended separately, just as molecular, electromagnetic, gravitational forces are in physics' (Lefebvre, 1991, p. 11). This chapter and ultimately this book has raised questions about … self-tracking though it has not answered them in full. There are treads to be picked up in future work. What I hope that I am leaving the reader with is the some but not all of the conceptual tools to inform critical inquiry, consumer awareness, design practice and to see not only analogies and connections between the goals and outcomes of self-tracking but also the breaks in the logic that brings goals and outcomes together. Self-tracking technology though it is becoming increasingly commonplace and fading into the background of our technically mediated lives is far from constituting a mature class of technology and the possibility for it to be developed in socially and politically progressive ways is something that we have a responsibility to carry forward through our respective disciplines.

References

Aarhus Institute of Advanced Studies. (2017). Metric culture: The quantified self and beyond. Retrieved from www.aias.au.dk/events/metric-culture

Adam, B. (1998). *Timescapes of modernity: The environment and invisible hazards*. London: Routledge.

Adler, P. S. (1993). Time-and-motion regained. *Harvard Business Review*, *71*(1), 97–108.

Ajana, B. (2013). *Governing through biometrics: The biopolitics of identity*. Basingstoke: Palgrave Macmillan.

Akrich, M. (1992). The de-scription of technical objects. In W. Bijker & J. Law (Eds.), *Shaping technology/building society* (pp. 205–224). Cambridge, MA: MIT Press.

Albrechtslund, A., & Lauritsen, P. (2013). Spaces of everyday surveillance: Unfolding an analytical concept of participation. *Geoforum*, *49*, 310–316. doi:10.1016/j.geoforum.2013.04.016

Alessi, S. M., Barnett, N. P., & Petry, N. M. (2017). Experiences with SCRAMx alcohol monitoring technology in 100 alcohol treatment outpatients. *Drug and Alcohol Dependence*, *178*, 417–424.

Anjum, R. L. (2016). Evidence based or person centered? An ontological debate. *European Journal for Person Centered Healthcare*, *4*(2), 421–429.

Apperely, T. (2011). *Gaming rhythms: Play and counterplay from the situated to the global*. Amsterdam: Institute of Networked Cultures.

Armstrong, A., & Hagel, J. (2000). The real value of online communities. *Knowledge and Communities*, *74*(3), 85–95.

Arnold, M. (2003). On the phenomenology of technology: The "Janus faces" of mobile phones. *Information and Organisation*, *13*(4), 231–256.

Austin, J., & Hardyman, P. (1991). The use of early parole with EM to control prison overcrowding: Evaluation of the Oklahoma department of corrections pre parole supervised release with EM. (National Institute of Justice).

Ayobi, A. (2018, April). Informing the design of personal informatics technologies for unpredictable chronic conditions. In *Extended abstracts of the 2018 CHI conference on human factors in computing systems* (pp. 1–4). New York, NY: ACM.

Ayobi, A., Sonne, T., Marshall, P., & Cox, A. L. (2018, April). Flexible and mindful self-tracking: Design implications from paper bullet journals. In *Proceedings of the 2018 CHI conference on human factors in computing systems* (pp. 1–14). New York, NY: ACM.

Bachelard, G. (1950). *La Dialectique de la Durée*. Paris: Presses Universitaires de France.

Bandura, A. (1982). Self-efficacy mechanism in human agency. *American Psychologist*, *37*(2), 122–147.

Bauer, J. J., McAdams, D. P., & Pals, J. L. (2008). Narrative identity and eudaimonic well-being. *Journal of Happiness Studies*, *9*(1), 81–104.

Bauman, Z. (2000). *Liquid modernity*. Hoboken, NJ: John Wiley & Sons.

Baumer, T. L., & Mendelsohn, R. I. (1991). Comparing methods of monitoring home detention: The results of a field experiment. Meeting of the American Society of Criminology (San Francisco).

BCC Research. (2017). Health self monitoring: Technologies and global markets (HLC185B). Retrieved from https://www.bccresearch.com/market-research/healthcare/health-self-monitoring-technologies-markets-report.html

Becker, G. S. (2007). Health as human capital: Synthesis and extensions. *Oxford Economic Papers*, *59*(3), 379–410.

Bendelow, G. A., & Williams, S. J. (1998). *The lived body: Sociological themes, embodied issues*. London: Routledge.

Boesel, W. E. (2013). What is the quantified self now? *The Society Pages*. https://thesocietypages.org/cyborgology/2013/05/22/what-is-the-quantified-self-now/

Boesel, W. E. (2017). My numbers sucked but I made this baby anyway. Quantified self. Retrieved from https://quantifiedself.com/show-and-tell/?project=1079

Bossewitch, J., & Sinnreich, A. (2013). The end of forgetting: Strategic agency beyond the panopticon. *New Media & Society*, *15*(2), 224–242.

boyd, D., & Crawford, K. (2012). Critical questions for big data: Provocations for a cultural, technological, and scholarly phenomenon. *Information, Communication & Society*, *15*(5), 662–679.

Brand, S. (1968). Purpose, function. *Whole Earth Catalog*. Retrieved from https://monoskop.org/images/0/09/Brand_Stewart_Whole_Earth_Catalog_Fall_1968.pdf

Brand, S. (1988). Editorial. *Signal*.

Brand, S., & Crandall, R. E. (1988). The media lab: Inventing the future at MIT. *Computers in Physics*, *2*(1), 91–92.

Brannan, M., Pearson, G., & Worthington, F. (2007). Ethnographies of work and the work of ethnography. *Ethnography*, *8*(4), 395–402.

Bratton, B. H. (2016). *The stack: On software and sovereignty*. Massachusetts, MA: MIT Press.

Brey, P. (2008). Technological design as an evolutionary process. In P. E. Vermaas, P. A. Kroes, A. Light, & S. Moore (Eds.), *Philosophy and design* (pp. 61–75). Dordrecht; London: Springer.

Brûlé, D. A. (2015). *Wearable RFID storage devices*. U.S. Patent Application No. 14/408,740.

Buchanan, R. (1992). Wicked problems in design thinking. *Design Issues*, *8*(2), 5–21.

Burrell, W. D., & Gable, R. S. (2008). From BF skinner to Spiderman to Martha Stewart: The past, present and future of electronic monitoring of offenders. *Journal of Offender Rehabilitation*, *46*(3–4), 101–118.

Burrows, R., & Nettleton, S. (2002). Reflexive modernization and the emergence of wired self-help. In R. Pea (Ed.), *Building virtual communities: Learning and change in cyberspace* (pp. 249–268). Cambridge: Cambridge University Press.

Bush, V. (1945). As we may think. *The Atlantic Monthly*, *176*(1), 101–108.

Calvard, T. (2019). Integrating social scientific perspectives on the quantified employee self. *Social Sciences*, *8*(9), 262. Retrieved from https://www.mdpi.com/2076-0760/8/9/262.

Carmichael, A. (2010, April, 5). Why I stopped tracking. The Quantified Self. Retrieved from https://quantifiedself.com/blog/why-i-stopped-tracking/

Carrigan, M. (2014, July 31). Self-Tracking and social control: What would techno-fascism look like? Retrieved from https://markcarrigan.net/2014/07/31/self-tracking-and-social-control/

Cirillo, F. (2006). The Pomodoro technique. In *Agile processes in software engineering and extreme programming*. Retrieved from http://www.baomee.info/pdf/technique/1.pdf

Clough, P. T. (2008). The affective turn: Political economy, biomedia and bodies. *Theory, Culture & Society*, *25*(1), 1–22.

Cooke, R. A., & Rousseau, D. M. (1988). Behavioral norms and expectations: A quantitative approach to the assessment of organizational culture. *Group & Organization Studies*, *13*(3), 245–273.

Coovert, M. D., & Thompson, L. F. (Eds.). (2013). *The psychology of workplace technology*. Abingdon: Routledge.

Couldry, N. (2004). Media meta-capital: Extending the range of bourdieu's field theory. In D. L. Swartz & V. L. Zolberg (Eds.), *After bourdieu: Influence, critique, elaboration* (pp. 165–189). Dordrecht: Springer.

Couldry, N., & Mejias, U. A. (2019). Data colonialism: Rethinking big data's relation to the contemporary subject. *Television & New Media*, *20*(4), 336–349.

Crabtree, A., Rodden, T., Tolmie, P., & Button, G. (2009, April). Ethnography considered harmful. In *Proceedings of the SIGCHI conference on human factors in computing Systems* (pp. 879–888). New York, NY: ACM Press.

Crane, E. (n.d.). Ellen Crane design (personal website). Retrieved from http://ellencranedesign.squarespace.com/strongarm-tech

Crawford, K., Lingel, J., & Karppi, T. (2015). Our metrics, ourselves: A hundred years of self-tracking from the weight scale to the wrist wearable device. *European Journal of Cultural Studies*, *18*(4–5), 479–496.

Cullen, F. T., & Gendreau, P. (1989). The effectiveness of correctional rehabilitation. In L. Goodstein & D. L. MacKenzie (Eds.), *The American prison* (pp. 23–44). Boston, MA: Springer.

Cullen, F. T., & Gendreau, P. (2000). Assessing correctional rehabilitation: Policy, practice, and prospects. *Criminal Justice*, *3*, 109–175.

Cyert, R. M., & Kumar, P. (1996). Economizing firms through learning and adaptation. *Journal of Economic Behavior & Organization*, *29*(2), 211–231.

Dalton, C. M., Taylor, L., & Thatcher, J. (2016). Critical data studies: A dialog on data and space. *Big Data & Society*, *3*(1), 1–9. doi:10.1177/2053951716648346

Dannehl, K. (2017). Object biographies: From production to consumption. In K. Harvey (Ed.), *History and material culture: A student's guide to approaching alternative sources* (pp. 171–186). New York, NY; London: Routledge.

Davis, J. L. (2020). *How artefacts afford: The power and politics of everyday things.* Cambridge, MA: MIT Press.

Deal, T. E., & Kennedy, A. A. (1982). *Corporate cultures: The rites and rituals of corporate life*. Reading, MA: Addison-Wesley.

Deloitte. (2019). Smart city, smart nation: Providing the keys to unlock your city's potential. Retrieved from www2.deloitte.com/content/dam/Deloitte/us/process-and-operations/us-cons-deloitte-smart-cities-trends-and-case-studies.pdf

Depedro, D., & Cannon, R. (1975). Electrocardiographic telemetry and telephone transmission link system. U.S. Patent No. 3,882,277. U.S. Patent and Trademark Office.

De Souza, R. N. A., Butt, D., Jethani, S., & Marmo, C. (2021). Participatory research methods for investigating digital health literacy in culturally and linguistically diverse communities. *Conjunctions. Transdisciplinary Journal of Cultural Participation*, *8*(1). doi:10.7146/tjcp.v8i1.117800

Diamandis, E. P. (2015). The hundred person wellness project and Google's baseline study: Medical revolution or unnecessary and potentially harmful over-testing? *BMC Medicine*, *13*(1), 5. doi:10.1186/s12916-014-0239-6

van Dijk, T. A. (2006). Ideology and discourse analysis. *Journal of Political Ideologies*, *11*(2), 115–140.

Dilts, A. (2010). *From 'entrepreneur of the self' to 'care of the self': Neoliberal governmentality and Foucault's ethics*. Western Political Science Association 2010 Annual Meeting Paper. Retrieved from https://ssrn.com/abstract=1580709

Dolejsova, M., Kera, D., Storni, C., Khot, R. A., Clement, I. J., Pavelka, I., & Kishor, P. (2017, May). Digital health & self-experimentation: Design challenges & provocations. In *Proceedings of the 2017 CHI conference extended abstracts on human factors in computing systems* (pp. 510–517). New York, NY: ACM Press.

Dunne, A., & Raby, F. (2001). *Design noir: The secret life of electronic objects*. Springer Science & Business Media.

Duval, S., & Wicklund, R. A. (1972). *A theory of objective self awareness*. Cambridge, MA: Academic Press.

Easterling, K. (2018). *Medium design*. Strelka Press.

Edensor, T. (2011). Commuter: Mobility, rhythm and commuting. In T. Cresswell & P. Merriman (Eds.), *Geographies of mobilities: Practices, spaces, subjects* (pp. 189–204). Farnham: Ashgate.

Everything 2. (2001). Schwitzgebel machine. Retrieved from https://everything2.com/title/Schwitzgebel+Machine

Fairs, M. (2014, March 10). Wearable technology that ignores emotional needs is a major error: Interview with Gadi Amit. *Dezeen*. Retrieved from https://www.dezeen.com/2014/03/10/interview-fitbit-designer-gadi-amit-wearable-technology/

Fallan, K. (2008). De-scribing design: Appropriating script analysis to design history. *Design Issues*, *24*(4), 61–75.

Featherstone, M. (1982). The body in consumer culture. *Theory, Culture & Society*, *1*(2), 18–33.

Fisher, T. H. (2004). What we touch, touches us: Materials, affects, and affordances. *Design Issues*, *20*(4), 20–31.

Fitabase. (2016). Fitbit and fitabase reach research milestone. Retrieved from https://www.fitabase.com/blog/post/fitbit-and-fitabase-reach-research-milestone/

Fitbit. (2016). How is Fitbit calculating calorie burn goal and why isn't it adjusting? Retrieved from https://community.fitbit.com/t5/One/How-is-Fitbit-calculating-calorie-burn-goal-and-why-isn-t-it-adjusting/td-p/1463586

Flanagin, A. J., & Waldeck, J. H. (2004). Technology use and organizational newcomer socialization. *Journal of Business Communication*, *41*(2), 137–165.

Fombrun, C. J. (1983). Corporate culture, environment, and strategy. *Human Resource Management*, *22*(1–2), 139–152.

Ford, H., & Graham, M. (2016). Provenance, power and place: Linked data and opaque digital geographies. *Environment and Planning D: Society and Space*, *34*(6), 957–970.

Forlizzi, J., DiSalvo, C., Bardzell, J., Koskinen, I., & Wensveen, S. (2011). Quality control: A panel on the critique and criticism of design research. In *CHI'11 extended abstracts on human factors in computing systems* (pp. 823–826). doi: 10.1145/1979742.1979497

Foucault, M. (1982). The political technology of individuals. In L. H. Martin, H. Gutman, & P. H. Hutton (Eds.), *Technologies of the self: A seminar with Michel Foucault* (pp. 145–162). Amherst, MA: The University of Massachusetts Press.

Foucault, M., Davidson, A. I., & Burchell, G. (2008). *The birth of biopolitics: Lectures at the collège de France, 1978–1979*. Berlin: Springer.

Fox, R. G. (1987). Dr Schwitzgebel's machine revisited: Electronic monitoring of offenders. *Australian and New Zealand Journal of Criminology*, *20*(3), 131–147.

Franssen, M. (2008). Design, use, and the physical and intentional aspects of technical artifacts. In P. A. Kroes, A. Light, & S. Moore (Eds.), *Philosophy and design philosophy and design* (pp. 21–35). Dordrecht: Springer.

Gable, R. K. (1986). Application of personal telemonitoring to current problems in corrections. *Journal of Criminal Justice*, *14*(2), 167–176.

Gable, R. K., & Gable, R. S. (2005, June). Electronic monitoring: Positive intervention strategies. *Federal Probation*, *69*, 21.

Gaver, B., Dunne, T., & Pacenti, E. (1999). Design: Cultural probes. *Interactions*, *6*(1), 21–29.

Gawne, M. (2012). The modulation and ordering of affect: From emotion recognition technology to the critique of class composition. *Fibreculture Journal*, *21*, 98–123. Exploring Affective Interactions.

Gerhard, U., & Hepp, A. (2018). Appropriating digital traces of self-quantification: Contextualizing pragmatic and enthusiast self-trackers. *International Journal of Communication*, *12*, 682–700.

Giddens, A. (1991). *Modernity and self-identity: Self and society in the late modern age*. Cambridge: Stanford University Press.

Gitelman, L. (Ed.). (2013). *Raw data is an oxymoron*. MIT press.

Goldenfein, J. (2015, July 23). What does the panopticon mean in the age of digital surveillance? *The Guardian*. Retrieved from https://www.theguardian.com/technology/2015/jul/23/panopticon-digital-surveillance-jeremy-bentham

Gram, M. (2019). On design thinking. n+1. 35. Retrieved from https://nplusonemag.com/issue-35/reviews/on-design-thinking/

Grant, J. (2010). Foucault and the logic of dialectics. *Contemporary Political Theory*, *9*(2), 220–238.

Haddon, L., & Kommonen, K. H. (2003). *Interdisciplinary explorations: A dialogue between a sociologist and a design group*. Chimera Working Paper Number: 2004-03. University of Art and Design, Helsinki.

Hamann, T. H. (2009). Neoliberalism, governmentality, and ethics. *Foucault Studies*, *6*(February), 37–59. doi:10.22439/fs.v0i0.2471

Haraway, D. J. (1997). The virtual speculum in the new world order. *Feminist Review*, *55*(1), 22–72.

Harris, S. G. (1994). Organizational culture and individual sense-making: A schema-based perspective. *Organization Science*, *5*(3), 309–321.

Harvey, D. (1972). Revolutionary and counter-revolutionary theory in geography and the problem of ghetto formation. *Geographic Thought*, *4*(2), 13.

Hassan, R. (2011). *The age of distraction: Reading, writing, and politics in a high-speed networked economy*. Transaction Publishers.

Hatano, Y. (1993). Use of the pedometer for promoting daily walking exercise. *ICHPER*, *29*, 4–8.

Hepp, A. (2016). Pioneer communities: Collective actors in deep mediatisation. *Media, Culture & Society*, *38*(6), 918–933.

Hepp, A. (2020). The fragility of curating a pioneer community: Deep mediatization and the spread of the quantified self and maker movements. *International Journal of Cultural Studies*, *23*(6), 932–950. doi:10.1177/1367877920922867

Hepp, A., Alpen, S., & Simon, P. (2020). Beyond empowerment, experimentation and reasoning: The public discourse around the Quantified Self movement. *Communications*, *1* (ahead-of-print), 25–51. doi:10.1515/commun-2019-0189

Hepp, A., Breiter, A., & Friemel, T. N. (2018). Digital traces in context—An introduction. *International Journal of Communication*, *12*, 439–449.

Heyen, N. B. (2020). From self-tracking to self-expertise: The production of self-related knowledge by doing personal science. *Public Understanding of Science*, *29*(2), 124–138.

Hickey-Moody, A. C. (2020). New materialism, ethnography, and socially engaged practice: Space-time folds and the agency of matter. *Qualitative Inquiry*, *26*(7), 724–732.

Hirsch, J. A., James, P., Robinson, J. R., Eastman, K. M., Conley, K. D., Evenson, K. R., & Laden, F. (2014). Using MapMyFitness to place physical activity into a neighborhood context. *Frontiers in Public Health*, *2*(1). doi:10.3389/fpubh.2014.00019

Hitchcock, K. (2015, March). Dear life: On caring for the elderly. *Quarterly Essay*, *57*.

Hodson, R. (1998). Pride in task completion and organizational citizenship behaviour: Evidence from the ethnographic literature. *Work & Stress*, *12*(4), 307–321.

Hucklesby, A. (2009). Understanding offenders' compliance: A case study of electronically monitored curfew orders. *Journal of Law and Society*, *36*(2), 248–271.

Hurley, R. F., & Hult, G. T. M. (1998). Innovation, market orientation, and organizational learning: An integration and empirical examination. *Journal of Marketing*, *62*(3), 42–54.

Hyysalo, S. (2010). *Health technology development and use: From practice-bound imagination to evolving impacts*. New York, NY: Routledge.

Ihde, D. (2008). *Ironic technics*. Copenhagen: Automatic Press.

Iliadis, A., & Russo, F. (2016). Critical data studies: An introduction. *Big Data & Society*, *3*(2), 1–7. doi:10.177/2053951716674238

Irwin, A., & Michael, M. (2003). *Science, social theory and public knowledge*. New York, NY: McGraw-Hill.

Jewson, N. D. (1976). The disappearance of the sick-man from medical cosmology, 1770-1870. *Sociology*, *10*(2), 225–244.

Jones, J. C. (1992). *Design methods*. Hoboken, NJ: John Wiley & Sons.

Keating, P., & Cambrosio, A. (2003). *Biomedical platforms: Realigning the normal and the pathological in late-twentieth-century medicine*. Cambridge, MA: MIT Press.

Kelly, K. (2012). Quantified self San Francisco meetup #1. Retrieved from https://vimeo.com/49791232

Kendall, H. O., & Kendall, F. P. (1968). Developing and maintaining good posture. *Physical Therapy*, *48*(4), 319–336.

Kenward, M. G., & Jones, B. (2014). Crossover trials. In N. Balakrishnan (Ed.), *Methods and applications of statistics in clinical trials: Concepts, principles, trials, and design* (pp. 310–319). New York, NY: Wiley.

Khatchatourov, A., Chardel, P. A., Peries, G., & Feenberg, A. (2019). *Digital identities in tension: Between autonomy and control.* Newark, NJ: John Wiley & Sons.

Kilgore, J. (2014). The grey area of electronic monitoring in the USA: James Kilgore offers his personal insight into the political and economic forces driving expansion. *Criminal Justice Matters, 95*(1), 18–19.

Kim, T., Chang, A., Holland, L., & Pentland, A. S. (2008, November). Meeting mediator: Enhancing group collaboration using sociometric feedback. In *Proceedings of the 2008 ACM conference on computer supported cooperative work* (pp. 457–466). New York, NY: ACM Press.

Kipnis, A. B. (2008). Audit cultures: Neoliberal governmentality, socialist legacy, or technologies of governing? *American Ethnologist, 35*(2), 275–289.

Kitchin, R. (2014). Big data, new epistemologies and paradigm shifts. *Big Data & Society, 1*(1), 1–12.

Kofman, A. (2019, July 3). Digital jail: How electronic monitoring drives defendants into debt. *The New York Times.* Retrieved from https://www.nytimes.com/2019/07/03/magazine/digital-jail-surveillance.html

Kotter, J. P. (2008). *Corporate culture and performance.* New York, NY: Simon & Schuster.

Krippendorff, K. (1989). On the essential contexts of artifacts or on the proposition that “design is making sense (of things)”. *Design Issues, 5*(2), 9–39.

Krippendorff, K. (1990). Product semantics: A triangulation and four design theories. In V. Seppo (Ed.), *Product semantic '89.* Helsinki: University of Industrial Arts. Retrieved from http://repository.upenn.edu/asc_papers/254

Lanier, J. (2010). *You are not a gadget.* New York, NY: Random House.

Latour, B. (1992). Where are the missing masses? The sociology of a few mundane artefacts. In J. Law & W. E. Bijker (Eds.), *Shaping technology/building society: Studies in sociotechnical change* (pp. 225–258). Cambridge, MA: MIT Press.

Latour, B. (2005). *Reassembling the social: An introduction to actor-network theory.* Oxford: Oxford University Press.

Lazer, D., Pentland, A., Adamic, L., Aral, S., Barabasi, A. L., Brewer, D., … Jebara, T. (2009). Social science: Computational social science. *Science, 323*(5915), 721–723.

Leder, D. (1990a). Clinical interpretation: The hermeneutics of medicine. *Theoretical Medicine, 11*(1), 9–24.

Leder, D. (1990b). *The absent body.* Chicago, IL: University of Chicago Press.

Lefebvre, H. (1991). *The production of space.* Oxford; Cambridge, MA: Blackwell.

Lefebvre, H. (2013). *Rhythmanalysis: Space, time and everyday life.* New York, NY: Bloomsbury Publishing.

Lester, R. K., & Piore, M. J. (2004). *Innovation-the missing dimension.* Cambridge, MA: Harvard.

Letiche, H., & Lissack, M. (2009). Making room for affordances. *Emergence: Complexity and Organization, 11*(3), 61–72.

Levy, S. (1984). *Hackers: Heroes of the computer revolution.* Garden City, NY: Anchor Press.

Li, I., Dey, A., & Forlizzi, J. (2010, April). A stage-based model of personal informatics systems. In *Proceedings of the SIGCHI conference on human factors in computing systems* (pp. 557–566).

Lilly, J. R., & Ball, R. A. (1987). A brief history of home confinement and house arrest. *Northern Kentucky Law Review*, *13*, 343–374.

Lupton, D. (1992). Discourse analysis: A new methodology for understanding the ideologies of health and illness. *Australian Journal of Public Health*, *16*(2), 145–150.

Lupton, D. (2013). Quantifying the body: monitoring and measuring health in the age of mHealth technologies. *Critical Public Health*, *23*(4), 393–403.

Lupton, D. (2016a). *The quantified self*. Cambridge: Polity Press.

Lupton, D. (2016b). The diverse domains of quantified selves: Self-tracking modes and dataveillance. *Economy and Society*, *45*(1), 101–122.

Lupton, D. (2018). How do data come to matter? Living and becoming with personal data. *Big Data & Society*, *5*(2), 1–11. doi:10.177/2053951718786314

Lyon, D. (2001). *Surveillance society: Monitoring everyday life*. Buckingham: Open University Press.

Lyon, D. (2013). *The electronic eye: The rise of surveillance society — computers and social control in context*. Hoboken, NJ: Wiley.

Lyon, D. (2018). *What is rhythmanalysis?* London; New York, NY: Bloomsbury Publishing.

Mandel, L., Ong, A. M., & Singer, P. (1975). Ambulatory patient monitoring system. U.S. Patent No. 3,898,984. U.S. Patent and Trademark Office.

Mann, S., & Ferenbok, J. (2013). New media and the power politics of sousveillance in a surveillance-dominated world. *Surveillance and Society*, *11*(1/2), 18–34.

Mann, S., Nolan, J., & Wellman, B. (2003). Sousveillance: Inventing and using wearable computing devices for data collection in surveillance environments. *Surveillance and Society*, *1*(3), 331–355.

Manwaring, K., & Clarke, R. (2015). Surfing the third wave of computing: A framework for research into eObjects. *Computer Law & Security Review*, *31*(5), 586–603.

Marcuse, H. (1941). Some social implications of modern technology. *Zeitschrift für Sozialforschung*, *9*(3), 414–439.

McKillop, M., Mamykina, L., & Elhadad, N. (2018, April). Designing in the dark: Eliciting self-tracking dimensions for understanding enigmatic disease. In *Proceedings of the 2018 CHI conference on human factors in computing systems* (pp. 1–15). New York, NY: ACM Press.

McLoughlin, I. P., & Clark, J. (1988). *Technological change at work*. Milton Keynes: Open University Press.

Michael, M., & Lupton, D. (2016). Toward a manifesto for the 'public understanding of big data'. *Public Understanding of Science*, *25*(1), 104–116.

Miettinen, K. (2008). Design: Structure, process, and function: A systems methodology perspective. In P. E. Vermaas, P. A. Kroes, A. Light, & S. Moore (Eds.), *Philosophy and design* (pp. 217–231). Dordrecht; London: Springer.

Mol, A. (2002). *The body multiple: Ontology in medical practice*. Durham, NC; London: Duke University Press.

Monogioudis, V. (2015, April 8). From counting steps to changing behaviour: Four design principles to build human centred, sticky and successful self-tracking services. *IDEO Stories*. Retrieved from https://medium.com/ideo-stories/from-counting-steps-to-changing-behaviour-cb8bebfdff24

Moore, P. (2018). Tracking affective labour for agility in the quantified workplace. *Body & Society*, *24*(3), 39–67.

Moore, P., & Robinson, A. (2016). The quantified self: What counts in the neoliberal workplace. *New Media & Society*, *18*(11), 2774–2792.

Morozov, E. (2013). *To save everything, click here: The folly of technological solutionism*. New York, NY: Public Affairs.

Munari, B. (1966). *Design as art*. London: Penguin.

Murphie, A. (2014). Auditland. *Portal Journal of Multidisciplinary International Studies*, *11*(2). doi:10.5130/portal.v11i2.3407

Nafus, D., & Sherman, J. (2014). This one does not go up to 11: The quantified self movement as an alternative big data practice. *International Journal of Communication*, *8*, 1784–1794.

Nagy, P., & Neff, G. (2015). Imagined affordance: Reconstructing a keyword for communication theory. *Social Media + Society*, *1*(2), 1–9. doi:10.1177/2056305115603385

National Institutes of Health. (2015). Precision medicine initiative cohort program Retrieved from https://www.nih.gov/sites/default/files/research-training/initiatives/pmi/pmi-working-group-report-20150917-2.pdf

Nettleton, S. (2009). Commentary: The appearance of new medical cosmologies and the Re-appearance of sick and healthy men and women: A comment on the merits of social theorizing. *International Journal of Epidemiology*, *38*(3), 633–636.

Nicolson, M. (2009). Commentary: Nicholas Jewson and the disappearance of the sick man from medical cosmology, 1770–1870. *International Journal of Epidemiology*, *38*(3), 639–642.

Nissenbaum, H., & Brunton, F. (2013). Political and ethical perspectives on data obfuscation. In M. Hildebrandt, & K. de Vries (Eds.), *Privacy, due process and the computational turn: The philosophy of law meets the philosophy of technology* (pp. 171–195). London: Routledge.

Noble, S. U., & Roberts, S. T. (2016). Through Google-colored glass (es): Design, emotion, class, and wearables as commodity and control. In S. Tettegah & S. Noble (Eds.), *Emotions, technology, and design* (pp. 187–212). San Diego, CA: Academic Press.

Norman, A. (1978). Project goldfish. *Information Privacy*, *1*(1), 8–15.

Norman, D. A. (1999). Affordance, conventions, and design. *Interactions*, *6*(3), 38–43.

Norman, D. A. (2008). The way I see IT signifiers, not affordances. *Interactions*, *15*(6), 18–19.

Nudelman, G. (2016, March 1). What does your wearable UX sound like? Retrieved from https://medium.com/@designcaffeine/ux-synesthesia-imagine-your-wearable-iot-ux-as-a-piece-of-music-8a324dc09c1d#.ehlvch9ls

Oliphant, P. C., & Derrick, B. G. (2015). Tracking device incorporating cuff with cut resistant materials. U.S. Patent No. 9,129,504. Patent and Trademark Office.

Orlikowski, W. J., & Gash, D. C. (1994). Technological frames: Making sense of information technology in organizations. *ACM Transactions on Information Systems*, *12*(2), 174–207.

O'Neill, C. (2017). Taylorism, the European science of work, and the quantified self at work. *Science, Technology & Human Values*, *42*(4), 600–621.

Pantzar, M., & Ruckenstein, M. (2015). The heart of everyday analytics: Emotional, material and practical extensions in self-tracking market. *Consumption Markets & Culture*, *18*(1), 92–109.

Pedersen, I. (2005). Mobility, human-centricity, and the design of wearable augmented reality interfaces. *International Journal of the Humanities*, *3*(1), 143–154.

Pentland, A. (2014). *Social physics: How good ideas spread—The lessons from a new science*. New York, NY: Penguin.

PEW Charitable Trusts. (2016, September 7). Use of electronic offender-tracking devices expands sharply. Retrieved from https://www.pewtrusts.org/en/research-and-analysis/issue-briefs/2016/09/use-of-electronic-offender-tracking-devices-expands-sharply

Pigrum, D. (2012). Semiotics, creativity and international education. In J. Thompson & M. Hayden (Eds.), *International Education* (p. 119). Milton Park: Routledge.

Piras, E. M. (2019). Beyond self-tracking: Exploring and unpacking four emerging labels of patient data work. *Health Informatics Journal*, *25*(3), 598–607.

Prior, L. (2003). *Using documents in social research*. London: SAGE Publications.

Prior, L. (2009). Commentary: From sick men and women, to patients, and thence to clients and consumers—The structuring of the 'patient' in the modern world. *International Journal of Epidemiology*, *38*(3), 637–639.

Protzen, J. P., & Harris, D. J. (2010). *The universe of design: Horst Rittel's theories of design and planning*. London: Routledge.

Pruitt, J., & Grudin, J. (2003, June). Personas: practice and theory. In *Proceedings of the 2003 conference on Designing for user experiences* (pp. 1–15).

Quantified Self. (2012). Kevin Kelly on the history and future of the quantified self Retrieved from https://quantifiedself.com/blog/kevin-kelly-on-the-history-and-future-of-qs/

Quantified Self. (n.d.). Meet ups. Retrieved from https://quantifiedself.com/meetups/

Quora. (2016). What is it like to live under house arrest? Retrieved from https://www.quora.com/What-is-it-like-to-live-under-house-arrest

Ramirez, E. (2013, August 17). Matthew Ames on stepping on the scale. The quantified self. Retrieved from https://quantifiedself.com/blog/matthew-ames-on-stepping-on-the-scale/

Ramirez, E. (2014, June 18). Brian cain on optimising productivity. The quantified self. Retrieved from https://quantifiedself.com/blog/brian-crain-optimizing-productivity/

Ramirez, E. (2015). Work with us. The quantified self. Retrieved from https://quantifiedself.com/blog/work-us/

Reddit. (2011). IAMA guy who wore the SCRAM bracelet for one year. Retrieved from https://www.reddit.com/r/IAmA/comments/hib13/iama_guy_who_wore_the_scram_bracelet_for_one_year/

Reddit. (2017). r/showerthoughts. Retrieved from https://www.reddit.com/r/Showerthoughts/comments/5vrdrv/a_fitbit_is_like_a_house_arrest_bracelet_that/

Reio, T. G., Jr, & Sutton, F. C. (2006). Employer assessment of work-related competencies and workplace adaptation. *Human Resource Development Quarterly*, *17*(3), 305–324.

Renzema, M., & Mayo-Wilson, E. (2005). Can electronic monitoring reduce crime for moderate to high-risk offenders? *Journal of Experimental Criminology*, *1*(2), 215–237.

Renzi, A. (2008). The space of tactical media. In M. Boler (Ed.), *Digital media and democracy: Tactics in hard times* (pp. 71–100). London: MIT Press.

Rich, E., & Miah (2017). Mobile, wearable and ingestible health technologies: Towards a critical research agenda. *Health Sociology Review*, *26*(1), 84–97.

Riskin, J. (2003). The defecating duck, or, the ambiguous origins of artificial life. *Critical Inquiry*, *29*(4), 599–633.

Roberts, S. (2010). The unreasonable effectiveness of my self-experimentation. *Medical Hypotheses*, *75*(6), 482–489.

Robinson, G., & McNeill, F. (2008). Exploring the dynamics of compliance with community penalties. *Theoretical Criminology*, *12*(4), 431–449.

Rothenberg, D. (1995). *Hand's end: Technology and the limits of nature*. Chicago, IL: University of California Press.

Ruckenstein, M., & Pantzar, M. (2017). Beyond the quantified self: Thematic exploration of a dataistic paradigm. *New Media & Society*, *19*(3), 401–418.

Ruckenstein, M., & Schüll (2017). The datafication of health. *Annual Review of Anthropology*, *46*, 261–278.

Ruhleder, K., & Jordan, B. (1997). Capturing complex, distributed activities: Video-based interaction analysis as a component of workplace ethnography. In A. S. Lee & J. Liebenau (Eds.), *Information systems and qualitative research* (pp. 246–275). Boston, MA: Springer.

Ryff, C. D. (1989). Happiness is everything, or is it? Explorations on the meaning of psychological well-being. *Journal of Personality and Social Psychology*, *57*(6), 1069.

Ryff, C. D. (2017). Eudaimonic well-being, inequality, and health: Recent findings and future directions. *International Review of Economics*, *64*(2), 159–178.

Ryff, C. D., & Singer, B. H. (2006). Best news yet on the six-factor model of well-being. *Social Science Research*, *35*(4), 1103–1119.

Santini, S., & Romer, K. (2006. June). An adaptive strategy for quality-based data reduction in wireless sensor networks. In *Proceedings of the 3rd international conference on networked sensing systems (INSS 2006)* (pp. 29–36). TRF.

Sassen, S. (2008). *Territory, authority, rights: From medieval to global assemblages*. Princeton, NJ: Princeton University Press.

Sauvanet, P. (2000). *Le Rythme et la Raison, t. II: Rythmanalyses*. Kimé.

Schenwar, M. (2014). *Locked down, locked out: Why prison doesn't work and how we can do better*. Oakland, CA: Berrett-Koehler Publishers.

Schenwar, M., & Jan, T. (2015). The quiet horrors of house arrest, electronic monitoring, and other alternative forms of incarceration. *Mother Jones*. Retrieved from https://www.motherjones.com/politics/2015/01/house-arrest-surveillance-state-prisons/

Schifani, A. M. (2020). *Urban ecology and intervention in the 21st century Americas: Verticality, catastrophe, and the mediated city*. London: Routledge.

Schmidt, A., & Kunzmann, C. (2006. October). Towards a human resource development ontology for combining competence management and technology-enhanced workplace learning. In *OTM confederated international conferences on the move to meaningful internet systems* (pp. 1078–1087). Berlin: Springer.

Schön, D., & Argyris, C. (1996). *Organizational learning II: Theory, method and practice*. Reading, MA: Addison-Wesley Publishing Company.

Schreiber, A. (2016. July). A provenance model for quantified self data. In *International conference on universal access in human-computer interaction* (pp. 382–393). Springer.

Schüll, N. D. (2016). Data for life: Wearable technology and the design of self-care. *BioSocieties, 11*(3), 317–333.

Schwitzgebel, R., Schwitzgebel, R., Pahnke, W. N., & Hurd, W. S. (1964). A program of research in behavioral electronics. *Behavioral Science, 9*(3), 233–238.

Schwitzgebel, R. K. (1972). Limitations on the coercive treatment of offenders. *Criminal Law Bulletin, 8*(4), 267–320.

Schwitzgebel, R. L., & Schwitzgebel, R. K. (1974). *Psychotechnology: Electronic control of mind and behaviour*. New York, NY: Holt Rinehart & Winston.

Sharon, T., & Zandbergen, D. (2017). From data fetishism to quantifying selves: Self-tracking practices and the other values of data. *New Media & Society, 19*(11), 1695–1709.

Sherman, L. W., Gottfredson, D., MacKenzie, D., Eck, J., Reuter, P., & Bushway, S. (1997). *Preventing crime: What works, what doesn't, and what's promising? A report to the United States congress*. Washington, DC: National Institute of Justice.

Simpson, P. (2012). Apprehending everyday rhythms: Rhythmanalysis, time-lapse photography, and the space-times of street performance. *Cultural Geographies, 19*(4), 423–445.

Snow, D. A., Rochford, E. B., Jr, Worden, S. K., & Benford, R. D. (1986). Frame alignment processes, micromobilization, and movement participation. *American Sociological Review, 51*(4), 464–481.

Spreenberg, P., Salomon, G., & Joe, P. (1995, May). Interaction design at IDEO product development. In *Conference Companion on Human Factors in Computing Systems* (pp. 164–165).

Stephens, M. (2018, August 14). Leah heiss on how human centred design methodology can create social impact. *Impact Boom*. Retrieved from https://www.impactboom.org/blog/2018/8/14/leah-heiss-on-how-human-centred-design-methodology-can-create-social-impact

Stephenson, A., McDonough, S. M., Murphy, M. H., Nugent, C. D., & Mair, J. L. (2017). Using computer, mobile and wearable technology enhanced interventions to reduce sedentary behaviour: A systematic review and meta-analysis. *International Journal of Behavioral Nutrition and Physical Activity, 14*(1), 1–17. doi:10.1186/s12966-017-0561-4

Strathern, M. (2000). New accountabilities. In M. Strathern (Ed.), *Audit cultures: Anthropological studies in accountability, ethics and the academy* (pp. 1–18), London: Routledge.

Sullivan, S. (2016). *Designing for wearables: Effective UX for current and future devices*. Beijing: O'Reilly Media.

Swan, M. (2009). Emerging patient-driven health care models: An examination of health social networks, consumer personalized medicine and quantified self-tracking. *International Journal of Environmental Research and Public Health, 6*(2), 492–525.

Swan, M. (2012). Sensor mania! The internet of things, wearable computing, objective metrics, and the quantified self 2.0. *Journal of Sensor and Actuator Networks, 1*(3), 217–253.

Swan, M. (2013). The quantified self: Fundamental disruption in big data science and biological discovery. *Big Data, 1*(2), 85–99.

Teli, M., Di Fiore, A., & D'Andrea, V. (2017). Computing and the common: A case of participatory design with think tanks. *CoDesign, 13*(2), 83–95.

Thacker, E. (2004). *Biomedia*. Minneapolis, MN: University of Minnesota Press.

Till, C. (2019). Creating "automatic subjects": Corporate wellness and self-tracking. *Health*, *23*(4), 418–435.

Tiwari, R. (2010). *Space, body, ritual: Performativity in the city*. Lanham, MD: Lexington Books.

Tolich, M. (1992). Bringing sociological concepts into focus in the classroom with "modern times," "Roger and me," and "Annie Hall". *Teaching Sociology*, *20*(4), 344–347.

Topol, E. (2011). *The creative destruction of medicine: How the digital revolution will create better health care*. Paris: Hachette.

Tudor-Locke, C., Hatano, Y., Pangrazi, R. P., & Kang, M. (2008). Revisiting "how many steps are enough?". *Medicine & Science in Sports & Exercise*, *40*(7), S537–S543.

Turner, B. S. (1984). *The body and social theory*. New York, NY: Blackwell.

Verbeek, P. P. (2006). Materializing morality: Design ethics and technological mediation. *Science, Technology & Human Values*, *31*(3), 361–380.

Verbeek, P. P. (2008). Morality in design: Design ethics and the morality of technological artifacts. In P. E. Vermaas, P. A. Kroes, A. Light, & S. Moore (Eds.), *Philosophy and design* (pp. 61–75). Dordrecht; London: Springer.

Waber, B. (2013). *People analytics: How social sensing technology will transform business and what it tells us about the future of work*. Upper Saddle River, NJ: FT Press.

Want, R., Hopper, A., Falcao, V., & Gibbons, J. (1992). The active badge location system. *ACM Transactions on Information Systems*, *10*(1), 91–102.

Wendt, T. (2015). *Design for Dasein: Understanding the design of experiences*. San Bernardino, CA: Thomas Wendt.

Whitson, J. R. (2013). Gaming the quantified self. *Surveillance and Society*, *11*(1/2), 136–176.

Wicklund, R. A., & Eckert, M. (1992). *The self-knower: A hero under control*. Boston, MA: Springer.

Williams, R. (1979). *Politics and letters: Interviews with New Left Review*. London: NLB.

Williams, R. (2014). *Keywords: A vocabulary of culture and society*. Oxford; New York, NY: Oxford University Press.

Winner, L. (1980). Do artifacts have politics? *Daedalus*, *109*(1), 121–136.

Winston, B. (1986). *Misunderstanding media*. Cambridge, MA: Harvard University Press.

Wolf, G. (2007, November 12). What is self-efficacy? The quantified self. Retrieved from https://quantifiedself.com/blog/what-is-selfefficacy/

Wolf, G. (2008, September 12). *Ping's thesis — From diary to graph*. The quantified self. Retrieved from https://quantifiedself.com/blog/from-diary-to-graph/

Wolf, G. (2009, August, 22). Know thyself : Tracking every facet of life from sleep to mood to pain 24/7/365. *Wired*. Retrieved from https://www.wired.com/2009/06/lbnp-knowthyself/

Wolf, G. (2010a). *The quantified self*. TED. Retrieved from https://www.ted.com/talks/gary_wolf_the_quantified_self/

Wolf, G. (2010b, April 28). The data-driven life. *The New York Times*. Retrieved from https://www.nytimes.com/2010/05/02/magazine/02self-measurement-t.html

Wolf, G. (2012). QS primer: Case crossover design. The quantified self. Retrieved from https://quantifiedself.com/blog/qs-primer-case-crossover-design/

Wolf, G. (n.d.). Retrieved from https://www.linkedin.com/in/aether

Wolf, G., & De Groot, M. (2020). A conceptual framework for personal science. *Frontiers of Computer Science*, *2*(21). doi:10.3389/fcomp.2020.00021

Wong, R. Y., & Khovanskaya, V. (2018). Speculative design in HCI: From corporate imaginations to critical orientations. In M. Filimowicz & V. Tzankova (Eds.), *New Directions in third wave human-computer interaction: Volume 2-methodologies* (pp. 175–202). Cham: Springer.

Wood, H. (2016). On san francisco's quantified self delusion. Retrieved from https://medium.com/@girlziplocked/on-san-francisco-s-quantified-self-delusion-ef5cc7adf95b

Woolgar, S. (1991). Configuring the user: The case of usability trials. In Law (Ed.), *A sociology of monsters—Essays on power, technology and domination* (pp. 57–99). London: Routledge.

You, H. C., & Chen, K. (2007). Applications of affordance and semantics in product design. *Design Studies*, *28*(1), 23–38.

Zeiler, K. (2010). A phenomenological analysis of bodily self-awareness in the experience of pain and pleasure: On dys-appearance and eu-appearance. *Medicine, Healthcare & Philosophy*, *13*(4), 333–342.

Zickar, M. J., & Carter, N. T. (2010). Reconnecting with the spirit of workplace ethnography: A historical review. *Organizational Research Methods*, *13*(2), 304–319.

Index

Printed and bound by CPI Group (UK) Ltd, Croydon, CR0 4YY

13/09/2026

14961061-0008